George F Harris

The Science of Brickmaking

With some account of the structure and physical properties of bricks

George F Harris

The Science of Brickmaking
With some account of the structure and physical properties of bricks

ISBN/EAN: 9783337035785

Printed in Europe, USA, Canada, Australia, Japan

Cover: Foto ©berggeist007 / pixelio.de

More available books at **www.hansebooks.com**

THE SCIENCE

OF

BRICKMAKING:

WITH SOME ACCOUNT OF

THE STRUCTURE AND PHYSICAL PROPERTIES OF BRICKS.

BY

GEORGE F. HARRIS, F.G.S.,

Membre de la Société Belge de Géologie, Paléontologie et Hydrologie; Lecturer on Geology, The Practical Applications of Geology, and Mineralogy, in the Birkbeck Institution, London; etc., etc., etc.

LONDON:

H. GREVILLE MONTGOMERY,

43, ESSEX STREET, STRAND.

1897.

PREFACE.

THE substance of this little work was first published as a series of articles in the *British Clayworker*, in 1895-96, and I am indebted to the courtesy of the Proprietor of that Journal for permission to reproduce them.

An attempt is here made to furnish some information of an elementary character on a special branch of technical education which has been seriously neglected in this country. But the reader will understand that it is only an elementary treatise. Its publication in serial form, where each article must, more or less, be complete in itself, has to a large extent determined the method of handling the subject, and I am fully cognisant of the drawbacks of the work in that respect.

At the same time, it is hoped that the book will be useful to the more advanced class of brickmakers and clayworkers generally, many of whom have expressed a desire to see the articles in this form.

GEO. F. HARRIS.

Birkbeck Institution,
    Bream's Buildings, Chancery Lane, E.C.
1st February, 1897.

CONTENTS.

THE SCIENCE

OF

BRICKMAKING.

CHAPTER I.

FLUVIATILE BRICK-EARTHS.

LET us go to Crayford, near Erith, or to Ilford, in Essex, and take a superficial glance at some of the brickyards found at those places ; in particular, let us ascertain a little concerning the earths employed. We find in one brickyard a series of stiff brown or bluish clays, interstratified between sandy clays or "loams,' with thin brownish partings. In another, the loam will become very sandy, and the earth light, with a slight greenish tinge. A third has thin pebble or gravel beds developed, or small stones sparingly scattered in the clays and loams on certain horizons. A fourth contains, in addition to some of the beds above described, a lime-clay or marl* with small pellets of chalk. It will be noticed on entering the yards that these various hori-

---

* This, and all other technical terms used, will be explained in an alphabetical glossary at the end of the book.

B

zons, or beds, as they are conveniently termed, are disposed in regular lines or layers, more or less horizontal; in other words, the beds are "stratified." On the face of the working being dug into, it will often be found that these thin beds, a few inches or feet each in thickness, vary in depth, and frequently disappear altogether, or "thin out," whilst, on the other hand, a bed only a few inches thick may become as many feet, and new beds are found to be developed. A pure sand may in like manner become loamy on being dug into, and, on being further developed, pass insensibly into a stiff clay. And many other changes take place into which we will not enquire at the moment. Suffice it to say, that in such brickyards the strata are very locally developed, though it follows from the circumstance of their existence for so many years, that what changes have taken place, to some extent compensate each other, so that the material is still an earth suitable for making bricks. Again, certain beds of much economic value may be more persistent than others, both in character and development. Having noticed all these things, we perceive a couple of men digging with care into the brick-earth, and presently they bring some objects to us which we have no difficulty in recognizing as the remains of the lower jaw of an elephant's skull. Returning to the spot where they were exhumed, the upper jaw and tusks also are uncovered. To the clayworkers these things are well known; in their time they have found many similar skulls of 'animals in the brick-earth; but they know next to nothing concerning them, or how they got there. Another expedition to the same localities may yield the remains of rhinoceros, the musk sheep, grizzly bear, hippopotamus, reindeer, and many other animals. A fine series of the remains of these, obtained from the

brick-earths of the valley of the Thames at several
points, is exhibited in the geological department of the
British Museum (Natural History), South Kensington,
and more or less complete skeletons obtained from the
same source may be found in other, and local museums.
One of the most interesting points concerning these
remains is that so many of the animals represented in
the brick-earths are of extinct species—there are no
species included in this latter category of precisely
similar kinds to animals now living Thus the elephant
was different to modern elephants ; we know, from
remains found elsewhere, that it was clothed with
wool. The same also with the rhinoceros. The reindeer no longer lives in this country, being confined to
northerly latitudes ; whilst the musk sheep is a deni-
zen of the Arctic regions, and the hippopotamus is
restricted to the tropical or sub-tropical climes. But
we might continue for a long time expatiating on the
character of the very numerous mammalian remains
found in our common brick-earths. What a curious
assemblage of animals ! It is wonderful to contemplate
the time when the reindeer and musk sheep lived side
by side with the elephant and rhinoceros on the site
whereon London now stands.

That is not all, however. In the same brick-earths
and gravels, tools (flint implements), fashioned by the
hands of man, are also frequently discovered, and in one
place at Crayford, the spot whereon flint implement
were manufactured has been 'lighted upon. Each flake
chipped off has been collected and pieced together, and
the shape of the original flint has thus been determined.
Clearly, from this evidence, the earth from which millions of bricks have been made has formed since
primæval man (and with him the animals alluded to)

inhabited the valleys of the Thames and its tributaries. It is interesting, too, to reflect on the circumstance that the materials upon which many of these facts of great philosophical significance are based, have been collected through the instrumentality of the workmen. Palæontologists are proud to acknowledge that ; their debt of gratitude to the intelligent and persevering men can never be fully repaid.

Pursuing the matter still further, we discover a quantity of shells, blanched and very frail—they seem to be deprived of much of their original substance, so to speak; their entombment in the brick-earth has taken all the natural colour out of them. Studying these, we soon ascertain that they belong to land snails and mollusca which inhabit fresh water. Living representatives of the same species are, with few exceptions, found in Kent and Essex.

Putting all this evidence together, we come to the conclusion that the brick-earths alluded to accumulated in the channel of a river ; they are found above the present level of the Thames, for the simple reason that they have been elevated into that position partly by earth movements and partly by the channel of the river being cut deeper by natural causes, of which abundant proof will be adduced. The snails were washed down from the land by freshets, or caught by the river in flood; the elephant, rhinoceros, hippopotamus, and musk sheep were overcome, perhaps, by floods, drowned, and subsequently covered up by the mud of the swollen current. We can imagine that the savage hunter, in his canoe, attacking the animals swimming in the river, loses his tomahawk, or his frail bark may be upset, and he is striving to gain the shore for dear life. Or, it may be winter time ; the river is frozen over, and he is cutting a hole in the ice with his flint chisel wherein to fish; his hands are benumbed, and he loses his grasp of the tool; it falls into the water, to be discovered in the brick-earth by one of our intelligent friends. Truly, the revelations of the brickyard enable us to construct a picture of one of the most interesting phases of the past history of the Earth.

We have given an outline of the evidence upon which certain brick-earths in the Thames valley are proved to be of fresh-water origin—to have accumulated in quiet reaches of the river, and at other convenient spots along its course—but we have used that as an illustration only; phenomena of precisely the same character are manifested in nearly all river valleys in this country, especially those in which the bottom of the valley has only a slight gradient down to the sea.

The brickmaker may ask: What is the practical bearing of these observations? What difference does it make to us whether the earths we use are of fresh-water, lacustrine, or marine origin? All the difference in the world, from the points of view of structure, composition and suitability of the earths, and especially of their distribution over the face of the country. How much easier it is to value an extensive brickmaking property when you feel perfectly certain as to whether the face of earth as shown in the pit will die out on being worked into for a few yards, or whether it will be persistent throughout the whole of the property to be valued. Better still, when your knowledge enables you to state definitely whether the quality of earth now being worked in a pit is likely to continue the same, or whether it will get better, or worse. The disposition of the earths, in some instances, is so clear that no brick-maker with an eye to business could fail to trace their extent over his property. But this is not often the case, for the earths being used are for the most part covered by a superficial mantle, or overburden, which masks the true character of the beds beneath. A very slight acquaintance with the principles of geology overcomes these difficulties as a rule; and we are about to lay down the elements of these principles, so far as they apply to the immediate subject in hand. By seeing why it is the beds of brick-earth vary in structure and composition we shall be in a better position to make forecasts of their general behaviour.

In regard to fluviatile deposits, it goes without saying that every river flows along a general depression more or less pronounced, called a valley, and that this valley is bounded physiographically by a ridge, except in the region of its entrance to the sea or lake, or, if a tributary, of its joining a main stream. The watershed of a river and its tributaries includes and comprises what is technically termed the "river basin." All valleys are, in the end, the result of denudation taking place in them. In other words, on the birth of a valley a very slight depression or other physical feature determined its general direction for the time being, but the little rivulet once being formed proceeded, through the medium of the "agents of denudation," to carve out its channel more clearly, and eventually to eat into the rocks over which it flowed, until a large valley had been formed. The "agents of denudation" in river valleys may be summarised as rain, snow, ice, heat, and wind, and their general effect on rocks is called "weathering." We need not stop to enquire into the precise methods adopted by these agents in accomplishing their work; it suffices at present to say that the rock destroyed or broken up is removed by the running water constituting the rivulet, stream, or river, as the case may be. Some of the material is chemically dissolved in the water, whilst another and larger proportion is taken away in suspension, or is said to be dealt with mechanically by the river. The agents of denudation do their work very slowly, as a rule, and yet no one who stands on London Bridge and contemplates the swollen stream laden with muddy sediment passing under it after a few days' rain, could say that they are not doing their duty effectually. To give some idea of the quantity of sand, gravel, and mud removed from the land through the medium of rivers, we may remark that the Mississippi discharges into the Gulf of Mexico annually a mass of earthy matter equal to a prism 268 feet in height with a base area of one square mile. In regard to denudation by chemical means we may say that the Thames carries past Kingston 19 grains of mineral salts in every gallon of water, or a total of 1,502 tons every 24 hours, or 548,230 tons every year; this is not taking into account the muddy sediment, gravel, &c., annually sent down to the Nore, which must be infinitely greater in quantity.

Enough has now been said to show that stupendous quantities of mineral matter derived from the destruction of the land are sent down to the sea by natural agencies, and we may at once state that a very large proportion of this, which finds a resting-place in and about the mouths of the rivers and their backwaters, is material suitable for brickmaking at places where it is obtainable. Enormous quantities of muddy sediment, sand and gravel, however, never reach as far as the sea with great rivers. This material is arrested at sundry convenient spots, and, as a rule, forms excellent brick-earth. See Fig. 1, which represents part of a river of slow

current with three bends, A, B, C. The water is flowing
in the direction indicated by the arrows; and it is part
of the mechanics of such a river that in rounding a
bend its velocity is greatest (and its eroding power also)
at the outer portions of the curves approximately
indicated by the arrow points. The water " wheels
round " such portions of the curves, and " marks time "
at the points *x x x*, and, indeed, its progress may be
altogether arrested for a time at the latter places. Now
the transporting power of a river is its velocity, and,
naturally, the greater the velocity, the coarser will be
the fragments or particles of rock carried along. It is
interesting in this connection to quote the figures cal-

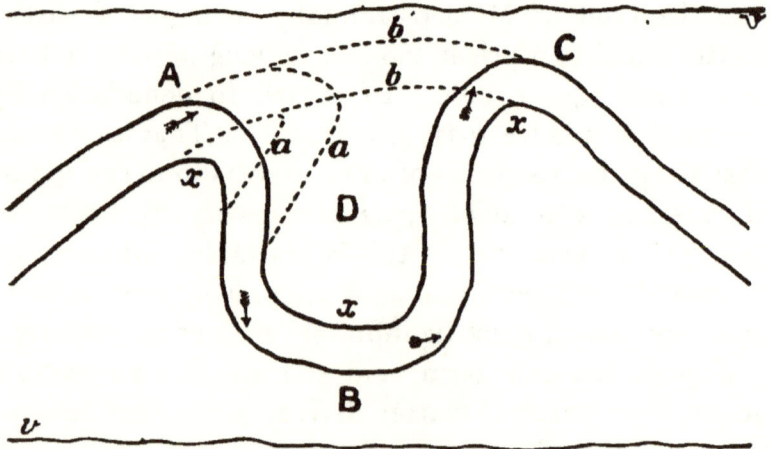

Fig. 1.—Formation of Brick-earth in a river valley.

culated by Mr. David Stephenson, giving the power
of transport of different velocities of river currents :—

Ins. per second.		Mile per hour.	
3	=	0·170	will just begin to work on fine clay.
6	=	0·340	will lift fine sand.
8	=	0·4545	will lift sand as coarse as linseed.
12	=	0·6819	will sweep along fine gravel.
24	=	1·3638	will roll along rounded pebbles 1 inch in diameter.
36	=	2·045	will sweep along slippery angular stones of the size of an egg.

These figures* have greater interest for us than in the connection at present used, as will be noticed hereafter. We have seen that in rounding the bends (Fig. 1) A, B, C, different portions of the stream possess different velocities. We know it is charged with sediment and stones all the time. The tendency, therefore, will be for the large stones and coarse detritus to go round the outer side of the bend, to bombard the banks near the points shown by the arrows, and to erode the channel deepest in those situations; whilst a goodly proportion of the fine muddy sediment will find its way to the quiet and shallow parts near *x x x*, and in course of time become deposited there, whilst the main course of the stream is eating its way and shifting its course as indicated by the dotted lines *a a*. This action proceeds, it may be, until the course of the river becomes straighter, as shown by the dotted lines *b b*, when the whole of the loop B D is abandoned, its former course there being evidenced by pools of water and irregular heaps of gravel, sand and mud. The reader will now see that the whole of the land marked *x x x* has been formed of sediment brought down by the river, and in the majority of cases such fine silt and sandy mud or clay is specially suitable for brickmaking—many of our largest brickmakers obtain their material from such a source. It should be observed that the valley, as shown between the lines *v v*, may be two or three miles in width, and it is often much more, so that the actual amount of land made by the river at *x x x* may be several thousands of acres in extent.

Now as to the practical application of the foregoing observations. In the first place, it will be seen that

---

* "Canal and River Engineering," p. 315.

such deposits of brick-earth as are made in this manner cannot be very thick, their total thickness perhaps, resting on the bottom of the valley, not being more than 20 feet, and it is frequently much less. The next thing to be noticed is that they must be very variable in character, a bed changing perhaps every 100 feet or so horizontally, and more often every few feet. Individual beds must of necessity be very irregularly developed under the circumstances. The velocity of the stream being greater at certain seasons of the year than at others, we frequently find some such section as the following developed :—

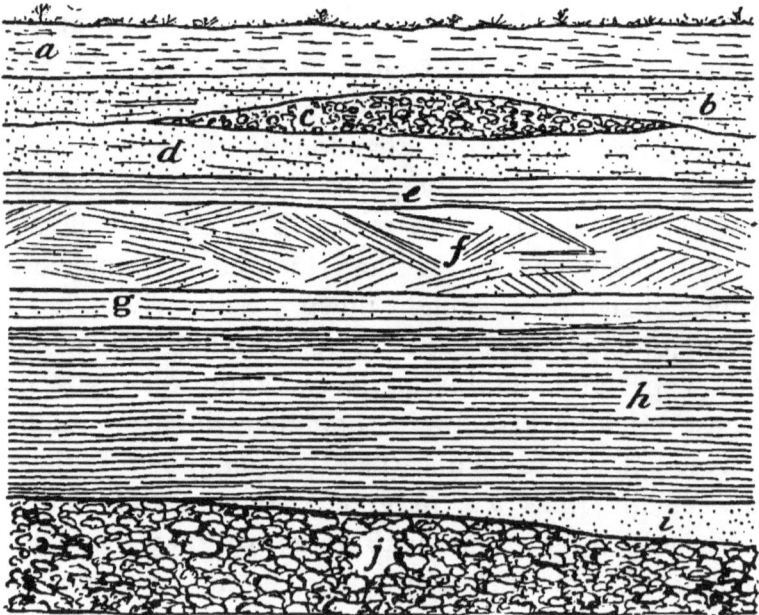

Fig. 2.—Section of Fluviatile Brick-earth

a = Mould and soil, of no use to the brickmaker.

b = Sandy clay, with a large proportion of sand; useful for moulding or incorporating with the "fat" clays below for brickmaking.

c = Gravel bed, lenticularly developed; suitable for mending roads, paths, &c.

d = Sandy clay; similar to b.

e = Thin bed of marl, with a fair proportion of lime.

f = Sands and small pebbles, irregularly stratified (false-bedded).

g = Laminated sandy clay.

h = Stiff clay; can be mixed with f and passed through the pug mill.

i = Sand; an irregular bed of very local occurrence.

j = Gravel bed, with much sand.

The above is typical of deposits accumulated in river valleys; it is different in character to deposits laid down in the sea (as will presently be described); the section exhibits very different classes of brick-earth also, and yields a totally different kind of brick to that obtainable from brick-earths of marine origin. The importance of the question of origin of a brick-earth, therefore, is just beginning to dawn upon us. Many rivers are noted as having throughout a long period of time wandered from one side of the valley to the other (by the process depicted in Fig. 1) several times, in which cases the brick-earth sections relating to them are liable to still greater variation. The reader would perhaps be very much astonished to find how much is known concerning peregrinations of that description in regard to particular localities, by competent authorities—usually field geologists.

We come to another important point in regard to river deposits. The ceaseless flow of the river, and the abrading action of the large stones rolled along at the bottom of its channel, tend to cut the latter deeper and deeper, and we have excellent evidence that most English rivers once flowed at a greater elevation in their valleys than they now do.   In consequence of this, the brickmaker may find his pit somewhat higher than the neighbouring river, which at an earlier stage of its existence made his brick-earths.   To a certain extent, small earth movements, as previously explained, are also undoubtedly responsible for many of these brick-earths now being at a considerable elevation above the surface of the river.   This phenomenon is illustrated in Fig. 3.

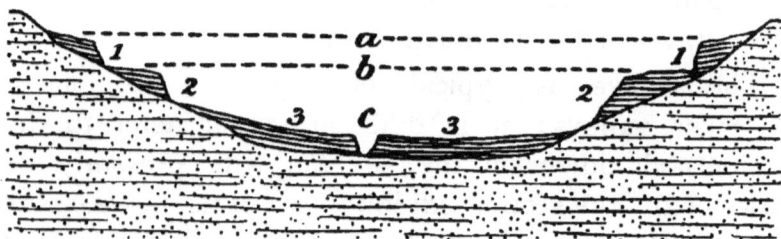

Fig. 3.—Section across a river valley, showing formation of terraces of gravel and brick-earth.

This type of disposition of fluviatile deposits is of common occurrence.   We will assume that the valley is carved out of clay (shown by horizontal lines and dots). On both sides of it, and at the same relative heights, are two masses (marked 1 and 2) of brick-earths and gravels running along so as to form two distinct broad terraces. These beds were laid down when the river, in flood, though occupying only a small portion of the valley, was approximately of the height shown by the dotted lines *a b*.   Denudation has been hard at work, however, since then, and only vestiges of these beds clinging to the sides of the valley, as shown, remain.   At a later period, and coming on towards modern times, the broad expanse of beds (comparable in disposition with those depicted in Fig. 2) some miles in width, marked 3, were laid down, and we notice the river channel, as it now is, cutting its way through them. Thus it comes to pass that brickyards may be situated in terraces one above the other ; and what is much more important, the brick-earths may vary very widely in quality along these horizons, those in 1 differing from 2 and both from 3. The brickyards may be quite close to each other, and to the unscientific eye the earths are of similar appearance, but they do not yield the same class of brick, and no one seems to trouble to enquire the reason why. These differences have resulted primarily from the materials having been derived from other collecting grounds, other watersheds, than those comprised within the basin of the river as at present constituted. They are the inevitable accompaniment of the evolution of the river system, and throw light on successive phases of the past history of the stream and its tributaries. For us, as we have seen, they possess considerable practical value of the first importance in selecting the site for a brickyard.

Apart from differences of the character just described,. serious alterations sometimes take place on these brick-earths being traced higher up the valley, and indeed an excellent brickmaking material may become absolutely worthless in that respect, for the reasons about to be explained. The reader will agree that neither stones nor sediment can travel up a valley, and he will understand that no sediment can be found in the valley earths other than that derived from the destruction of rocks within the watershed of the river system, to which the valleys belong, or did belong, at the time the earths were formed. We desire to put the case in a very simple light, so as to be clearly comprehended. Let us contemplate Fig 4. Here we have represented a

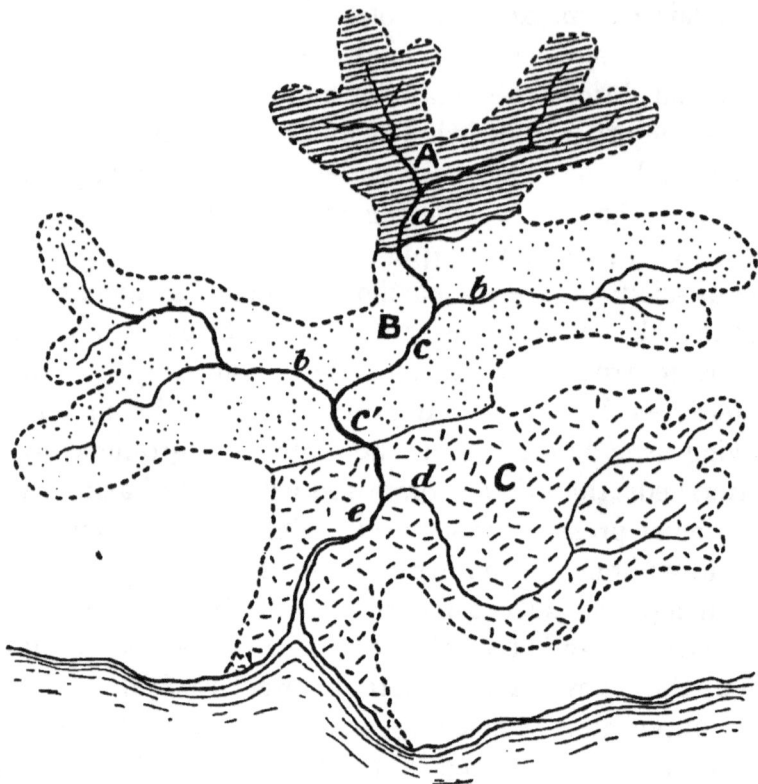

Fig. 4.—Map shewing river basin, with geological formations
depicted.

river basin, the limits (watershed) of which are indicated by a sinuous dotted line. Three geological formations are found therein ; in the upper reaches of the main
river is a series of clays marked A ; a large tract in the
middle, B, is sandstone ; and the lower part, C, is occupied
by limestone. Seeing that nothing but clay crops out
in the part A, it follows that the deposits of the river
in that region must be principally of an argillaceous
character, to the point a. On flowing over the sandstone B, the main stream, already charged with clay

particles, will be mixed with sand; the proportion of
sand increases as the first large tributary (*b*) to the east
is encountered, and is considerably augmented as the
still more important tributary (*b*) to the west enters it.
The superficial deposits in the valleys of the area B will
likewise be very sandy and perhaps gravelly at *b b*, but
at *c c*¹ the sands and gravels will be mixed with much
clay. On passing over into the area c, much carbonate
of lime is added, though the larger proportion denuded
from the rocks is taken away, chemically, in solution.
Nevertheless, nodules of "race" (lime concretions),
limestone pebbles, and perhaps chert and flint gravel
will come upon the scene at about the point marked *e*.
At *d* the deposits would principally consist of gravel
and impure marls. To sum up, the clays at *a* would
no doubt be too stiff of themselves to make good
bricks; similarly the beds at *b b* would be nothing
but sand, though these might be made, with a little
judicious treatment, into a species of fire-brick; at *c*
we should find alternating loams and clays suitable
for turning out fair bricks; at *c*¹ the beds would be
more variable in character and more locally developed;
they would consist of thin beds of sand, clays, loams and
gravels (principally sandstone fragments), which as a
whole might be made serviceable, though difficult to
deal with; nothing of much use to us would come
from point *d*, nor bordering the tributary running
over c; there would be too much lime present, though a
trade might be started in basic bricks should there be any
demand for them in the neighbourhood; this, however,
would only pay under extremely favourable conditions.
At *e* there may be a mixture of all the foregoing deposits, and providing the beds above were easily weathered
and thick beds of loam were thus fairly well developed,

good sites for brick-earth might be found. The point c might possess this advantage over the other sites mentioned, viz., that marls would no doubt be present, and thus no necessity should arise for grinding lime to be incorporated with the brick-earth ; the only danger would be that lumps of limestone might be too numerous—especially if c were a hard limestone.

The general character of the deposits might be slightly modified by mineral matter brought up in springs and thrown down at convenient spots.

CHAPTER II.

LACUSTRINE AND FLUVIATILE BRICK-EARTHS.

THE great variability of brick-earths deposited in river valleys is reflected to some extent in those laid down in lakes, though the size of the latter is frequently a controlling factor. The chief difference consists in the broader expanse of the sediment laid down—especially in large lakes—and variation in structure is not so noticeable horizontally. Let us consider a simple case in which a lake is fed by a large river bringing down abundant sediment. The lake acts as a species of settling tank, and the method of deposition of the sediment by the river is mainly guided by the velocity of the stream. The tendency under normal conditions is for the river to commence parting with its sediment immediately on entering the lake. The detritus alluded to is only held in suspension by the velocity of the water; when the latter is checked, as on entering the lake, the grosser pieces subside, and as its rapidity becomes progressively curbed, medium-sized fragments are compelled to give way, until at last only very minute particles are left in the water. In due time most of these also are deposited. Thus gravel is laid down before grit, grit before sand, and sand before clay.

If the velocity of the river always remained the same, we should be presented with thick accumulations of the same character in sharply defined areas. But it is always changing. With every storm and every steady

c

rain the motion of the river becomes greatly accelerated, with the result that the deposits for the time being are deposited farther out in the lake than in more quiescent periods. In this way we may have a gravel thrown down on sand, sand on clay, and so on.

From the foregoing observations it will be gleaned that, in general, deposits in large lakes are more persistent in character than are river deposits ; indeed, in very large sheets of water, as Lake Superior, Lake Erie, &c., they are in this sense more comparable with sediment of marine origin.

The practical value of this knowledge hinges on the correct determination of the origin of the deposits, and it is not always easy to identify a brick-earth of lacustrine origin. In all probability the tyro, on meeting one, would be disposed to regard it as a river deposit pure and simple. The valuation of a brick-earth property under such circumstances would thus be greatly in favour of the prospective purchaser ; but it would be disastrous for the seller. A random section, except in the case of a very large lake, would show gravels, sands and clays in much the same manner as the river deposits described in the last article of this series. But, as previously remarked, on the whole they would be more continuous and persistent, and what is quite as important, the mineral composition of each stratum would be equally homogeneous when traced over wide areas. The geologist distinguishes a lacustrine deposit from one of fluviatile origin more from its mineral constitution and the general disposition of the beds, as ascertained by mapping, than from evidence afforded by fossils—these latter for the most part being similar to those found in the deposits left by rivers.

The well-known brick-earth called " Reading mottled

clay," so extensively developed on the outskirts of the London basin, and in the Isle of Wight and Hampshire generally, furnishes a good example of a lacustrine deposit. Many millions of bricks are made from this bed every year, and in some parts of the districts mentioned the stratum is thick and extensively developed. It is pure enough to be suitable for terra-cotta manufacture here and there. No one who had seen this remarkable deposit could possibly fail to recognise it again. The natural colour of the clay when damp is brilliant red, scarlet or crimson, in large blotches and patches mottled tea-green and yellow, and locally white.

We have been intensely amused to note the efforts in recent years to obtain possession of a few acres of this coveted deposit for brickmaking in divers localities. Not long since we visited a large brickmaking establishment where these Reading plastic clays are actively raised and used, the works being situated four miles from the nearest railway. There were no other brickworks between it and the railway line, and there was no water accommodation. Enquiry revealed the fact that the greater part of the intervening land belonged to the same landowner as the ground where the brickyard stands, and that no difficulty was apprehended of the owner letting out such intervening land for the same uses and on the same terms if other brickyards were contemplated. The proprietor of the brickyard in question volunteered the information that the reason he started so far from the railway was because the earth at the point selected was the only kind suitable for brickmaking in the neighbourhood. We then questioned him as to his knowledge of the brick-earths in the district, and eventually elicited the fact that he chanced upon the spot selected, without any reasoning therefor,

and commenced operations. As a matter of fact, precisely the same clay extended from his works all the
way to the railway line, and had he known anything
whatever of the geology of the district (even the merest
boy's knowledge of the subject), he would have seen
how to save that four miles of road carriage. What
prevented him from knowing the fact was a thin mantle
of gravel and soil about four feet in thickness, which
covered the plastic clay in the area generally, except in
the immediate vicinity of his brickyard. That was in
reference to a lacustrine deposit—the Reading plastic
clay—and shows the value of knowing something of its
persistent character; if it had been a river deposit there
would not have been so much room for wonderment.

To give some idea of the extent of that particular horizon,
we may say that not only is the plastic clay alluded to found
so extensively in the London and Hampshire basins, it
is even more expanded in the north-eastern parts of
France, and is there as much utilised as on this side of
the Channel for brickmaking.

Lacustrine deposits are sometimes of enormous value
to the clayworker, on account of the general purity of
the clays. This is more particularly the case when the
material deposited is in part or wholly derived from
chemical disintegration of granitic rocks, as in the celebrated Bovey Heathfield clays near Newton Abbot, so
well described in a small pamphlet by Mr. S. Smith
Harvey. Here an experimental boring proved the clays
to a depth of 130 feet with no signs of exhaustion. In
the divers clay-pits but a small proportion of waste is
found, the different levels vary in composition, and,
like almost all thick clays, improve in quality as
the depth increases. The strata are very irregular
towards the surface, due perhaps to the action of local

freshets in the final periods of the history of the lake. These clays are extensively employed for the manufacture of stoneware pipes, facing and other bricks, fire-bricks, etc. They constitute a somewhat remarkable exception to the class of clays laid down in lakes, as a rule, and, as will have been observed, are of enormous thickness.

We have very little to say in regard to estuarine brick-earths; as might readily be anticipated, they are intermediate in character between fluviatile and marine deposits, and approach the one or the other according to position in the estuary. On the whole, they are variable in character, individual beds being thin. The strata frequently contain abundant plant remains (pieces of wood, etc.), and, except in the case of large rivers, are not noted for yielding very good brick-earths. Sometimes, however, the quality of the clays is not bad, as instance the bricks made in Lincolnshire and Northamptonshire from Jurassic Estuarine clays.

CHAPTER III.

MARINE BRICK-EARTHS.

TURNING to brick-earths of marine origin, we may say that these constitute by far the largest class of deposits from which bricks are made in this country, and it will be useful to deal with their origin in some detail. If we attentively watch the action of the weather on a friable sea-cliff we notice that large pieces tumble at intervals on to the beach, and in due time these are washed away by the waves, thus encouraging more to fall when the time is ripe. This process of denudation each year takes tens of thousands of tons of sandy clays and the like from the beaches around our islands. Large pieces of rock, too, are detached by the weather, and eventually succumb to wave action. During storms large stones are hurled against the cliffs, and the general effect of this bombardment is to wear them away, and reduce them to powder and sand grains with all possible expedition. No one who has not seen the waves at work at such times can have any idea of their tremendous power of moving blocks of stone many tons in weight. During calm weather the slight movement of the waves on the beach is manufacturing tons and tons of sand. A mass of gravel falls from the cliff; the finer particles are floated away at the earliest opportunity; the angular stones have their rough projections knocked off by striking against each other; and the incessant movement up and down the beach slope reduces the rough stone to a pebble,

all the time the particles thus shaved off are taken out to
sea for greater or less distances.   If the cliffs are of limestone, or similar rock, both chemical and mechanical
methods of denudation come into play, and considerable
quantities of lime, &c., are taken away by the sea water in
suspension and solution.   Large quantities of lime are
daily added to the sea through the agency of rivers also.

Now, what becomes of these vast quantities of detritus
furnished to the sea ?   That depends on the shore
currents at the particular locality.   If there is not much
of a current, the larger grains of grit and sand are soon
separated from the rest, and fall to the bottom, whilst the
clays are taken farther out to sea before being laid down.
But, in any case, the reader will readily perceive that
marine deposits must of necessity be on a grander scale,
and of a much more substantial character, as a rule, than
river, lacustrine, or estuarine deposits.   By their mode
of origin, too, they must be more homogeneous, whilst
they are frequently several hundreds of feet in thickness.
In their process of deposition they were not influenced by
every storm and freshet ; nothing short of great earth-
movements in process of time, or some other equally
grand phenomena, could disturb the even tenour of their
existence.   How different to the comparatively insignifi-
cant strata formed by the other methods alluded to !

Take samples of brick-earth of fluviatile origin at intervals and analyse them ; no two analyses will be alike,
except by a most remarkable coincidence—more by
accident than otherwise.   On the other hand, take a thick
marine clay, and compare its chemical composition as
ascertained at the present time with that of it made, say,
20 years ago in the same brickyard, and the analyses will,
in most instances, be practically identical—at any rate,
so far as they may be of use to the brickmaker.

A brickmaker using a marine clay possesses innumerable advantages over another employing brick-earths due to river action.   It is no uncommon thing for a marine clay—say, 300 feet in thickness—to continue across country for hundreds of miles, stretching from the North of England to the South, and over into the Continent, save for the slight break occasioned by the scooping out of the English Channel.   The composition of the Oxford Clay, from which the well-known bricks at Peterborough are made, does not differ in the slightest degree, so far as suitability for brickmaking is concerned, from the Oxford Clay of Bourges or Chateauroux, in the centre of France, or indeed at almost any other point *en route*.   With marine beds it is possible to deal with the matter on broad lines, but it is not so with any other class of deposits.

If a marine clay in a specified locality is found to be unsuitable for bricks at one point, by reason of the presence of too much lime, it would be a phenomenon if clay along the same geological horizon did not present the same unfavourable features at every other point within the district.   The homogeneous composition, both from mineralogical and chemical standpoints, of thick marine clays renders them of special use to the brickmaker.   Having by sundry processes, after infinite labour, produced a certain class of brick from such an earth, he does not as a rule have to materially modify those processes as the earth is dug into to continue manufacturing the same brick.   He is dealing with an earth which, comparatively speaking, is a constant quantity—when the clays are thick, and no lines of bedding are distinctly visible.

We find that a rooted conviction exists in many brickyards that clays of marine origin are no good for brickmaking, because (so the opinion runs) they always

contain so much salt. It is wonderful that such ignorance prevails, when the slightest acquaintance with the
subject would teach otherwise. It is perfectly true that
such deposits might have contained salt during and for
some time after deposition, but it is absurd to suppose
that their marine origin has anything to do with
the presence of common salt in the clay at the present
time. Salt is soluble in water, and has been removed
from such clays by the percolation of underground water
in 99 cases out of a hundred. Indeed, as a matter of
experience, we find that salt is most commonly found in
beds of lacustrine origin, or those laid down in enclosed
portions of the sea, for reasons we need not enter
into at the present moment. Of course, when material is
taken from the sea-shore to make into bricks, a considerable quantity of salt is manifest, but that is a
totally different thing to the clays deposited—we
should not like to say how many thousands of years ago.
Clays of all kinds, however, may be impregnated with
salt (as in parts of Cheshire), owing to the proximity of
other beds containing that mineral; also by the percolation of underground water with much salt in
solution.

To give some idea of the antiquity of the Oxford Clay
alluded to—and that is quite a " young clay " geologically speaking—we may remark that at the time it was
laid down not a single species of animal existed like
those now living. The only mammals found, very small
and very lowly organised, were like kangaroo rats; the
birds were more like flying reptiles than anything else; it
was the age of reptiles, and enormous, unwieldy brutes
swam in the water or floundered about on land; huge
sharks abounded, and armour-clad fish of kinds very
different to those now existing roamed the sea; even

the "shell-fish" were not altogether like modern
ones ; whilst the plants find their nearest modern ana-
logues in the wilds of Australasia.   No elephants, tigers,
lions, bears, or dogs lived then, and the face of
Nature wore a totally different aspect to what obtains
at the present time in any part of the globe.

And this seems a fitting opportunity to the writer to
put on record the fact that many of the most wonderful
remains found in the Oxford Clay and the neighbouring
Kimeridge Clay are due to the discoveries of brickmakers.   Without their valuable aid scientists would be
quite unable to clearly depict the life of those remote
epochs.   We have mentioned Peterborough ; some
most interesting remains have been found in the
clays near that town during the past few years.   To
appreciate this let the reader visit the fossil reptile gal-
lery of the British Museum (Natural History), at South
Kensington.   One of the most recent acquisitions, set
up a year or two ago, is the skeleton of a young *Plesiosaurus*—without doubt the most perfect specimen in the
world of its kind —from Peterborough.   The *Plesiosaurus*
was a large swimming reptile, with paddles, and a long
neck.

We mention these things not only to instil philosophical interest in such brick-earths, which may be reflected upon after business hours, but to impart some
idea of the extreme remoteness of the epoch from the
human point of view, and to insist on the immensity of
the intervening time throughout which circulating underground waters—even in such an impervious material as
stiff clay—may have exerted chemical action.   The
" mineralisation " of the fossils is an eloquent witness of
the effect of such changes.   The reader will perceive
from this that there is scant possibility of soluble salts

being present in such marine clays; and the geological circumstances are fully borne out by the results of hundreds of chemical analyses of thick marine clays.

The invertebrate fossils more particularly testify to the marine origin of the clays, and are thus invested with considerable practical interest. The man whose duty it is to determine the persistence, or otherwise, of valuable marine brick-earths has thus a much easier task than when called upon to decide the value of a large tract of land for brickmaking purposes, of fluviatile origin. Finally, brick-earths do not, except in extremely rare instances, vary materially in character when dug into horizontally, thus every opportunity is afforded to the manufacturer for making an unvariable quality brick, tile, or drain pipe. It should be borne in mind, however, that these clays often weather a brown colour, which on being dug into changes to a bluish-black tint, the latter being the unaltered and best portion as a rule. The only practical advantage the worker of a superficial river deposit possesses over his neighbour using thick marine clay is in the great range of variation in materials disclosed in the former kind of pit. By judiciously mixing the different beds he may be able to live well where the worker of marine clays, especially where the clay is too stiff, or contains too much lime, " comes to grief." A good marine clay is a great boon, a bad one cannot be remedied other than by the sacrifice of much money.

CHAPTER IV.

THE MINERAL CONSTITUTION OF BRICK-EARTHS.

THERE cannot be any question that the applicability or otherwise, of an earth for making good bricks, to a large extent depends on the mineral constitution of that earth. A chemical analysis of a sample of such earth will tell us how much silica. alumina, lime, iron, etc., is present therein, and this information is frequently of great value when given by a scientific chemist; but it does not tell us the state in which those constituents exist in the earth—an essential *desideratum*, if we are to understand the scientific aspects of the question of burning in the kiln. Further, the size of the granules and particles composing the earth is well worth knowing, as we shall presently see. It is a great mistake to imagine that all clays are essentially chemical deposits. The majority of them have been in part derived from chemical disintegration, it is true; but the resulting deposits contain so much also that is purely of mechanical origin, that the behaviour of the whole is materially modified, from a metallurgical point of view. Take one ingredient, for example—say, silica. That may exist in a brick-earth in a variety of ways, both in a free and combined state; but its behaviour in the kiln is largely dependent on the particular form assumed, not only whether it is free or combined, but as to how it is combined. In a certain sense, it is very doubtful whether even in the best-burnt brick much of the raw material becomes chemically combined ; a sort of agglutination takes place locally, as is clearly shown by the microscope ; at such points true fusion undoubtedly takes place, and there may be actual chemical combination.   In the vast majority of cases, however, such fusion or possible combination is of an extremely partial and elementary character, whilst it hardly exists in the average " rubber." The microscope shows that even in the hardest burnt brick there still remain enormous quantities of what may be termed mineral grains, that have by no means succumbed to the burning process.   The edges of the grains may occasionally be seen merging into the more or less vitreous ground mass in which they are embedded, but beyond that they appear tolerably fresh, and their action on polarised light remains unimpaired.

We did not intend to say anything yet concerning the microscopic structure of bricks—that will be gone into in a subsequent chapter; but we thought it useful to state the foregoing elementary facts in order to endeavour to uproot a conviction that seems to be very firmly grounded—viz., that the chemical composition of a brick-earth imparts an accurate idea of the possible active agents, on the earth being subjected to the kiln.   As a matter of fact, some of these would-be agents are imprisoned in the mineral grains and particles that have not become involved in the partial melting or agglutination of the mass, and might as well not be present in the earth for any work they may accomplish either for good or for evil.   There is greater probability of the bulk of these grains and particles being of active service when they are ground up exceedingly fine; but the clayworker's idea of " fineness," as demonstrated by what passes through an ordinary clayworking mill, and " fineness " in the sense here intended, are two totally different things.

We mean something that shall render the particles so small as that they shall only be observable on being magnified, say, 50 diameters. Hardly any clays used in brickmaking are in bulk made of such small particles as this; there are a few, of which the best terra-cotta and porcelain are manufactured, however, but even these have to be very carefully prepared to exclude grosser foreign particles. From what we have said, it will be gathered that the terra-cotta and porcelain manufacturer is at the present time in a better position to judge of the work done in the kiln or oven than is the brickmaker. But that is simply a matter of education; the problems presented to the average brickmaker are rather more complicated than to the terra-cotta manufacturer, but they may be unravelled on sufficient application, as we hope to point out.

Even under the most favourable conditions, however—when the particles composing the mass require a $\frac{1}{4}$-inch objective for their elucidation—we find that the best burnt brick is largely made up of them in an unmelted condition. And we should be very sorry to get rid of them; for if they disappeared, the stony attributes of the brick would disappear also, and the general value of the substance would be deteriorated to such an extent that it would be unsaleable as a building material. The brick would nearest resemble a form of slag. All we now insist upon is that in brickmaking a chemical analysis is only useful up to a certain point, beyond which we must appeal to the microscope to aid us, and this in conjunction with as perfect a knowledge as possible as to the behaviour of earths of certain mineral composition when under the influence of high temperatures. In many instances, the value of the brick depends almost entirely on incapacity for fusion on the part of a large proportion of the minerals of which the brick is made. Possibly, a good all-round brick would be where the bulk of its mineral particles were infusible at the temperature employed, and when the remainder were fusible enough to partially run, so as to cement or agglutinate the infusible particles firmly together. In order to bring about such conditions artificially, or to arrive at them even approximately, we must know at least three things, viz.—(1) the nature of the mineral particles involved in the whole operation ; (2) their behaviour under high temperatures ; and (3) a knowledge of certain branches of metallurgical chemistry. Now, obviously, we cannot undertake to teach even the spirit of what is involved in these three *desiderata* in a small book like this ; but we can, and shall, attempt to do something in that direction, and we must ask the reader's indulgence to take for granted observations to be occasionally made, in the inevitable prospect of our not being able to explain them at sufficient length.

The following are the principal minerals found in clays used in brickmaking, together with their more important attributes from our point of view.

KAOLIN.

Pure clay is, theoretically, composed of this mineral alone, but pure clay does not exist in Nature, except as a mineralogical curiosity. What is generally called pure clay is a white, or light-grey plastic material, composed of kaolin with many other substances to a small degree, from which it frequently has to, as far as possible, be separated before being put to its highest uses in porcelain manufacture. Chemically, pure kaolin may be regarded as a hydrous silicate of alumina, viz.—silica =

46·3, alumina = 39.8, and water = 13·9. Under the microscope, in reflected light, it is seen to be made up of extremely minute, thin, six-sided plates, which are said (doubtfully) to crystallize in the rhombic system ; though, when regarded with the naked eye, one would not suppose that it possessed a crystalline structure, as it appears to be an earthy, unctuous substance. It is commonly mixed with grains and small crystals and fragments of quartz, which mineral will presently be described. Being derived from the decomposition of felspars, the microscope reveals the fact that in addition to the six-sided plates alluded to, a great deal of opaque matter, as particles of mud, occurs in the substance universally known as kaolin. It is very difficult to satisfactorily state what this mud is ; micro-chemically, its general character may be brought out. There is no doubt, however, that in converting the kaolin into china-ware, these particles are more active than the minute kaolin crystals in uniting with other substances to form a species of flux. The subject has been investigated to a very limited extent, but from the foregoing observations it will be seen that the proportion of amorphous mud particles to the minute crystals must be an important factor in determining the nature of the fluxing material, and of the quantity of this latter to be used. Correlatively, the fusing point can be determined in the same manner. For, in itself, kaolin is an infusible mineral, and before it can be made use of for brick-making, terra-cotta, or any kindred purpose, it must be rendered artificially fusible by the addition of a fluxing substance. When, therefore, we learn that kaolin is being used for these purposes, we know, if used direct as it comes from the pit, that it must be impure from a mineralogical standpoint, or that it is being mixed with other substances. We say that kaolin is infusible (refractory) ; we mean at any temperature used in the industrial arts, including brickmaking. With the recent improvements in the electric furnace, the temperature generated is so high that practically any mineral sub-stance may be melted ; it is hard to speak of anything being infusible.

But the mineral matter called kaolin in ordinary clays, such as the brown and blue London Clay, the Oxford Clay, "brick-earths," etc., has very little in common with the more or less pure china-clay. The microscope shows that in the vast majority of such clays scales of true kaolin are few and far between, that opaque mud particles are more frequent, and, above all, that pieces of highly decomposed felspar (called " kaolinised " matter) are present. Eliminating all other and foreign substances from the clay, the whole of what would com-monly be called kaolin and kaolinised matter, taken to-gether, is of very varied chemical composition, and might, indeed, be fusible in the ordinary sense of that term. From this, the reader will perceive that the term kaolin is very ambiguous and altogether too wide in its mean-ing. We think it highly desirable, therefore, to des-cribe kaolin as a true mineral and not as a rock, reserving the term for the crystalline plates. The mud particles referred to we may call "kaolinised par-ticles ; " and the highly decomposed felspar "kaolinised matter." To sum up the relative fusi-bility of these substances, *per se*, we should say that (1) kaolin crystals are practically infusible ; (2) kaolin-ised particles are either fusible, partly fusible, or in-fusible, depending on the actual nature of the particles ; and (3) that kaolinised matter may be difficultly fusible or infusible. A mixture of (1) and (2) may not be

D

fusible, and could not be unless a great proportion of
(2) of a fusible character, so as to form a flux, were
present. The reasons for this will appear in consider-
ing the different kinds of felspar, next to be described.

FELSPAR.

This mineral, a very common constituent of nearly
all clays and brick-earths is very variable in character,
but may be separated into a number of mineral
species, each of which possesses a definite structure
and a more or less constant chemical composition.
To show the range of variation, the following kinds of
felspar, with their chemical composition, may be
quoted :—*

Chemical Composition of Felspars.

	Silica.	Alumina.	Potash.	Soda.	Lime.
Orthoclase	64·6	18·5	16·9	...	...
Albite	68 6	19·5	...	11·8	...
Oligoclase	63 /	23·9	1·20	8·1	2·0
Labradorite	52·9	30·3	...	4·5	12·3
Anorthite	43·0	36·8	...	...	20·1

Orthoclase felspar, in addition to the above, frequently
has small proportions of lime, iron, magnesia and soda.
Amongst other things it is an essential constituent of
granite, and on the decomposition of that rock is the first
mineral to become affected. When attacked in the open
air by rain and the ordinary agents of denudation, granite
ultimately gives way by the dissolution of the felspar,
and on being removed, the felspathic matter may accu-

---

* See, Geikie's " Text Book of Geology," 1882, p. 72.

mulate in convenient situations to form kaolin. If we now compare the chemical constitution of orthoclase felspar with that of kaolin as previously given, we notice that the potash has disappeared in the decomposing process; it has been dissolved and taken away by rivulets, and the like, or washed by rain direct into the sea. We also observe that there has been a re-distribution, so to speak, of the relative proportions of silica and alumina—following well-known laws.

Of the remaining felspars the commonest for our purposes is oligoclase, a mineral found in nearly all British "granites" in a greater or less degree. That contains a higher percentage of alumina than orthoclase, and there is a fair proportion of soda and little lime, but much less potash. The lime-soda felspar, labradorite, and its near ally, anorthite, are not often met with in a recognisable form in clays. If present, they are generally as "kaolinised matter," too highly decomposed to exhibit their characteristic optical properties.

It is pretty generally stated, and too often assumed by some, that pure china-clay is derived from the direct decomposition of rocks containing "orthoclase" felspar. Yet, this cannot really be so, if we reflect on the mineral composition of many of the rocks, which, obviously, have yielded the china-clays in question. Take the china-clays of Devon and Cornwall; they have undoubtedly been derived from the "granites" of those counties. To some extent, as previously remarked, the orthoclase is attacked, and provides the material of which china-clay is made. But in the "West of England" we have yet to learn that some of the other felspars are not also involved in the process. If we examine a fresh piece of granite from the flanks of Dartmoor, or from the neighbourhood of Liskeard, or

St. Austell, we find no difficulty in recognising a fair proportion of triclinic felspar (one or more of those mentioned in the table except orthoclase) in it. There is a difference in the composition (and therefore the commercial applicability) of a china-clay derived from a rock containing orthoclase alone, and one from a rock having orthoclase and one or more triclinic felspars in addition. The latter minerals are more easily decomposed than orthoclase, especially the lime and lime-soda varieties. We should not have raised this point only that, by reason of the granites being to some extent mechanically as well as chemically decomposed, a large proportion of "kaolinised particles" and "kaolinised matter" is introduced into certain china-clays, which render them different in their behaviour under intense heat from those china-clays in which orthoclase alone has been principally concerned. In other words, great practical advantages accrue from an accurate knowledge of the constitution and origin of the china-clays in question. Two clays of the same chemical composition often behave in a different manner in the kiln ; the cause of this is frequently to be found in the prevalence of "mechanical fragments" of felspar in one of the clays ; and the absence of these, but the presence of "kaolinised particles" of the same chemical composition, in the other.

Another point to which we may draw attention is the erroneous supposition that granites which have yielded china-clay have in all instances been reduced to the condition in which we now find them by the action of atmospheric agents of denudation alone. Granites, as a matter of fact, yield very slowly to the action of the atmosphere, and taken as a whole no building stone is as durable as they. How comes it, then, that they have decomposed to such an extent as to have formed extensive deposits of china-clay in a very short space of time, geologically? We think the answer is to be sought, at any rate in some instances, in the alteration the rock as a whole has undergone in certain situations, whereby it became more easily decomposable. Take the rotten china-stone of the neighbourhood of St. Austell, for example. In that material we clearly see a stone from which the "life" has been sapped, and instead of a bright, sparkling, porphyritic granite, as it once was, we now notice only a ghost of its former self. The large orthoclase felspars may be seen in it as skeletons, the mica is reduced to mere iron-stains (when present at all), whilst the quartz is also slightly affected. This altered and comparatively rotten material (although sometimes hard enough to be used as building stone) extends to an enormous depth from the surface; it has not been bottomed in some parts of the district. Such an extensive transformation could not possibly be due to ordinary agents of denudation which do their work at and near the surface of the rock only. It seems to arise from an enormous regional alteration, acting underground to an unfathomable depth, and which may not be unconnected with the mineral veins so common in, and in the immediate vicinity of the workings.*

Yet another thing to be remembered is that, under certain conditions, as near St. Austell, china-clay has been formed *in situ*, and has therefore not been deposited by the action of running water, as have the majority of china and other clays. Mr. Collins remarks that this china-clay is very irregular in its occurrence. It seems

---

* Information on this subject will be found in Mr. J. H. Collins' work, "The Hensbarrow Granite District." Truro, 1878.

to be formed of various granite masses decomposed in place; it often occupies considerable surface areas, and extends to a depth unknown. He remarks that at Beam mine, and also at Rocks mine, both near St. Austell, china-clay was found to a depth considerably exceeding 60 fathoms from the surface. This china-clay, in its natural condition, is very much the same as china-stone; but the decomposition has proceeded further, the felspar being completely changed into clay; and nothing more is necessary for extracting the clay than the disintegration of the whole mass by a stream of water directed upon it, when the clay is carried away in suspension and collected at convenient spots. Thus there is every gradation between the true crystalline orthoclase and triclinic felspars, through china-stone into china-clay formed *in situ*, so into china-clay deposited from water by natural or artificial means, and into a pure clay containing a large proportion of kaolin crystals, " kaolinised particles " and " kaolinised matter." But although we can state that much, a great deal yet remains to be done in connecting mineral structure with chemical composition of the purer clays, and in defining the various grades scientifically, in order that full advantage may be derived from them in a commercial sense.

CHAPTER V.

MINERALS: THEIR BEHAVIOUR IN THE KILN.

THE SILICA GROUP.

SILICA, the oxide of silicon, is found in brickmaking clays principally in two conditions when not combined with other substances: in one of these the free silica may be crystalline, when it is known as *quartz*; in the other it may be hard, but not crystalline, as *flint*. We may consider these in order.

QUARTZ.—When pure this mineral is perfectly white and transparent, like ordinary window glass. It is exceedingly hard, and this property is of much service as enabling us by the most elementary examination to distinguish it from certain other minerals, which it is not unlike at first sight. One of the latter is calcite, a crystalline form of carbonate of lime, also white and transparent. Quartz and calcite behave in a very different manner in the kiln, and as we shall see, they are both rather common constituents of brick-earth. The difference in hardness may easily be ascertained by the point of a good steel knife; the steel will not scratch the quartz, but it will, easily, the calcite.

When it has plenty of room wherein to crystallise, and is not hemmed in, as it were, by other hard crystalline matter, quartz often forms beautiful six-sided prisms surmounted, by a six-sided pyramid, and, rarely, pyramids are found at both ends of a prism. There are no lines, or "planes of cleavage," to interfere with the transparency, either in the extremely minute forms of the mineral as investigated by the microscope, or in the gigantic crystals occasionally found. Regular crystals of quartz, although by no means rare in Nature, are seldom met with entire in brick-earths. The most common form of the mineral is in irregular aggregates with other minerals, as in the rock granite, which is composed essentially, as previously mentioned, of quartz, felspar, and mica. We have traced the history of the felspar on the decomposition of that rock, and it may now be said that on complete disintegration of the granite a great part of the quartz present is simply resolved into fragments and dealt with by rain and other transporting agents. For quartz is practically imperishable; it is almost proof against the deleterious acids in the atmosphere, which so readily attack many other common minerals. In dealing with it, all Nature can do (at least at the surface of the earth) is to carry the small quartz grains and pieces about from place to place; She can, and does, in this process, reduce the quartzose fragments by causing them to continually knock against each other and against other mineral fragments and masses until the grains and pieces find a resting place; She may put them in a mill and grind them to powder, but the quartz is still there.

Another manner in which quartz occurs in Nature is as filling cracks in rocks, but this is comparatively unimportant for our present purposes. The purest quartz is known as rock crystal; but by far the commonest kinds of the mineral are impure; they may contain iron, schorl (a black needle-like crystal), and many other minerals. One of the most interesting points about it, and which undoubtedly in certain cases is of importance to the brick manufacturer as modifying its melting properties, is the presence of myriads of extremely minute so-called cavities, generally filled (or nearly filled) by liquids of different kinds, the precise nature of which is not as well-known as it might be, though in some instances it has been determined, with tolerable certainty.   In some cases these inclusions are so numerous as to obliterate the transparency of the quartz crystal, causing it to present a frosted appearance.   The fluids in these cavities may have beautiful little crystals of other minerals, such as salt, floating about—but it must be remembered that we are referring to something infinitely little.   These slight differences in the constitution of minerals, however, have their influence in the kiln.   For instance, although the fluid present is usually water, that often contains carbon dioxide, which acts as a species of flux to the quartz when present in sufficient quantity.

In reference to the second form of silica present in brick-earths, flint, that is of precisely the same chemical composition as quartz, only that it is not crystalline, nor transparent, though thin pieces of flint are translucent. Flint is by no means as common in Nature as quartz ; it is very hard, but brittle, and breaks with what is termed a conchoidal fracture, from the fact that the fractured surface frequently resembles the external appearance of the shell of a bivalve mollusc.   It occurs in a variety of ways ; (1) often as hard lumps or nodules running along in fairly regular layers in limestone rocks such as chalk, and (2) occasionally filling up cracks or joints in such rocks.   It is hard to describe its origin in a few words, and we shall not attempt it ; all that need be noted is that it is frequently full of the remains of extinct organisms of small size, which may, or may not, constitute an impurity depending on the particular organism and its present condition.   When flint contains a fair proportion of iron it is called chert—an extremely common constituent of brick-earths in some localities—though that term refers to other rocks, such for instance, as those made up almost exclusively of the siliceous spicules (hard parts made of silica) of fossil sponges.

A more or less crystalline kind of silica is found, forming the skeletons of minute aquatic plants, and these accumulating to some depth, constitute the basis of such materials as Kieselguhr and the diatom earth of the Isle of Skye, both of which, especially the former, are used for making firebricks.

There is very little to be said concerning the behaviour of free silica—quartz and flint—in the kiln. It is infusible except at higher temperatures than are employed by the brickmaker. But, as we have already remarked, the impurities often present in the minerals form a species of flux which naturally brings them into the range of fusible substances, though even then the temperature required is far beyond what is usually attained in the majority of brickyards, though it might be frequently arrived at in the manufacture of certain fire-bricks. For all ordinary purposes, therefore, quartz and flint may be regarded as infusible. In presence of much lime, iron, or similar substances, however, both of them are readily melted, and it is part of the science of brickmaking to know exactly how much lime, &c., to add to yield the best results. Many brick-earths contain large quantities of the calcareous and ferruginous substances alluded to, and are then capable of being made into bricks direct, without any addition. But although such natural brickmaking earths are frequently employed by the manufacturer, nearly all of them could be made to yield a better brick by a little artificial mixing. We must keep urging this point; there is room for great improvement all round.

As with the majority of comparatively refractory substances, the size of the grains and pieces of quartz and
flint makes a difference in their readiness to become
fusible. The larger the grain the more difficult it is to
break down ; fusion commences at the outside of a quartz
grain, the centre of which may at the same time be comparatively unaffected. By arresting the fusing process,
the microscope shows the outside of the grain to have
become softened (so much so as to affect its doubly re-
fracting properties), whilst the innermost parts still retain
their usual optical characters.

MICA

The different varieties of mica are important as rock-
forming minerals, but they are not as often met with in
brick-earths as is generally supposed, except in insignifi-
cant quantity. Some of the purest clays, however, contain a great deal of mica, derived almost directly from
the destruction of granite. The two commonest varieties
of the mineral are *biotite* and *muscovite*.

BIOTITE MICA.—This mineral, usually known as
ferro-magnesian mica, is composed of silicates of magnesia, alumina, iron, and alkalies in variable proportions.
It occurs as six-sided plates or irregular scales, usually
of a bronze-black colour. Biotite weathers with comparative facility, hence the reason why it is not more commonly met with in brown and other impure clays.

MUSCOVITE MICA.—This is sometimes called potash-
or alumino-alkaline mica, composed of the silicates of
alumina, alkalies, iron, and magnesia ; the proportion
of silica ranges from 45 to 50 per cent. It may usually
be distinguished at sight from biotite by its silvery white
or light brown colour. When large enough, both the

micas mentioned may be split up into thin plates, muscovite yielding large transparent sheets. Compared with
all other constituents of brick-earth, the micas are bright
and of semi-metallic lustre. Muscovite is more durable
than biotite, and is much more frequently met with in
brick-earths, especially in the sandy varieties.

The influence of mica in the kiln is not of much importance in ordinary brickmaking; in general its
alkaline character renders it fusible, though a high temperature is necessary at all times to effect that. In china-
clay mica is regarded as a nuisance, and in breaking down
the material it is separated in the washing process by
running water, the mineral collecting in depressions or
basins, called " micas." When muscovite contains
much fluorine, as it frequently does, it is very undesirable in clays for high-class purposes. At the best of
times the proportion of iron in mica is sufficient to mar
the quality of the otherwise most excellent clays. In the
kiln, or porcelain furnace, the presence of mica (more
particularly biotite) is apt to create yellow and brown
specks, or a species of mottling. It is highly satisfactory, therefore, to note that these little shiny flakes may
be easily floated off by a moderate amount of care in
washing, and thus separated from the other constituents
of the clay.

IRON.

Except in regard to white kaolin clays, nearly all
earths used in brickmaking contain more or less iron,
which is usually present as protoxide in many mineral
constituents. The colouring matter of clays is generally iron in some form, and blue clays weather into
brown by the alteration of that mineral. It is unnecessary for us to consider the various minerals of the iron

group; all we need do is to state the mode of occurrence of iron oxides in clays and earths, to consider a variety known as iron-pyrite, and the general effects of ferruginous minerals in the kiln.

Iron may occur in clays simply as a stain, when it is usually not in large quantity, or it may occur combined with some mineral or minerals present—as for instance certain felspars and micas. The brown, yellow, or blue appearance of the clay is due to it. In loam it may be found also as a species of ochreous earth, and in thin bedded loams (as the upper part of the Woolwich and Reading series of the London basin) each layer frequently varies in the proportion of iron present. In the more arenaceous parts of these loamy deposits, little grains of iron sometimes make their appearance, as also in certain sands employed in brickmaking; on careful examination, however, many of these grains are found to be other mineral substances coated with iron. Certain horizons in what are known as the Jurassic rocks contain great quantities of ferruginous matter in little pellets.

Iron, in large proportion, acts as a flux to other constituents when the brick-earth is subjected to great heat in the kiln, and on that account must be carefully watched. But, to the average brickmaker, the ferruginous constituent is far more interesting as a colouring medium. At a later stage we shall have something to say concerning the colouring of bricks, &c., but it may now be remarked that red bricks, in practically all cases, owe their colour to the effects of firing on iron. It is a great mistake to imagine, however, that a large percentage of iron in a clay will necessarily produce a good red tint. In the first place, a great deal depends on the way the clay has been mixed or prepared; and in the second, the method of burning and the temperature employed, taken in conjunction with the general composition of the earth, are all important. This much may be said, however, that without the iron (or some mineral colouring matter possessing similar properties in the kiln) a red brick would not result. An even colour is the effect of thorough and homogeneous incorporation of the iron with the brick-earth ; that may have been brought about by natural processes, but it is most frequently obtained in the careful prepartion and mixing of the clays. A very essential point is that the earths must be of such a character as to withstand the requisite heat in the kiln without becoming vitreous, or twisting or warping. It must not be forgotten that a certain proportion of the iron, under great temperatures, may be carried away out of the kiln in union with other things, in the form of vapour. To successfully treat a raw earth, so that all these points may be taken into account, and to produce a thoroughly uniform red brick, that shall not vary in tint from kiln to kiln, is a matter requiring considerable skill and attention, though fairly good bricks of that character have been produced by sheer accident in burning natural earths fairly rich in thoroughly dissemi-nated iron oxides.

Two minerals commonly met with in earths used for brickmaking are pyrite and marcasite, both of which are of the same chemical composition, namely, iron disulphide. We may first consider them separately, for they are of great importance to the brickmaker.

Iron pyrite occurs as regular cubic crystals, or irregular streaks, or as nodules or lumps ; in clay, the last-mentioned is its commonest form. It is a good petrifying medium, so that it is frequently associated with organic remains, as is exemplified in almost any yard where stiff clay is being worked.   The nodules, on being broken open, ordinarily exhibit a radiating structure of brassy lustre and extremely beautiful appearance, though often marred by brown iron stains due to decomposition of the mineral.   In the refuse of slates, now so largely used in several parts of the world for brickmaking, pyrite is most frequently found as fine cubic crystals of a durable nature.

Marcasite, on the other hand, crystallizes in a different manner (in the rhombic system of mineralogists), but is chiefly found in fibrous masses or dirty-brown nodules, the last-mentioned being common in clays.   When bright it is paler in tint than pyrite, though this is not a constant character.   It occurs abundantly in almost all sedimentary rocks diffused as minute particles, but sometimes in irregular layers.   Sir Archibald Geikie states * that this form of the sulphide is especially characteristic of stratified rocks, and more particularly of those of Secondary and Tertiary age.   That it is not abundant in Primary rocks is not to be wondered at when we consider its liability to rapid decomposition ; indeed, for it to be preserved at all it must be well shielded from atmospheric agents by Nature.   Exposure even for a short time to the air causes it to become brown, free sulphuric acid is produced, which may attack surrounding minerals, sometimes at once forming sulphates, at other times decomposing aluminous silicates and dissolving them in considerable quantity.   It plays even a larger part than pyrite as a petrifying medium, at any rate in the younger rocks.   Both pyrite and marcasite are abundant in many other rocks than those of special interest to the brickmaker ; the former, in fact, is almost universal in its occurrence.

* " Text Book of Geology," 1882, p. 85.

It will be convenient to consider the behaviour of these two minerals in the kiln together, as the difference between them from that point of view is practically *nil*. Under the action of the intense heat met with there, they become partially decomposed; oxide of iron and basic sulphides of iron remain. When, at a subsequent period, bricks containing these substances are exposed to the action of the weather, oxidation takes place, sulphate of iron and sometimes of lime are formed, which on crystallizing expand with considerable force and split or crack the brick. From this it is evident that sulphide of iron in any form is not to be tolerated in brick manufacture, and if the earth used in the first place contains much, it must be removed in the preparing process. If permitted to remain, it is impossible to obtain either a durable, or a good coloured brick.

CHAPTER VI.

MINERALS: THEIR BEHAVIOUR IN THE KILN *(continued).*

CALCITE, ARAGONITE, &c.

CARBONATE of lime may occur in a crystalline form, or as earthy substances, and many varieties of it are found n clays used by the brickmaker. The commonest are calcite, aragonite, and a white earth.

Calcite, known also as calc-spar, crystallises in the hexagonal system, though true hexagons are not very common. It occurs principally as rhombohedra and scalenohedra, with variations therefrom; also fibrous, lamellar, granular, compact, nodular, and stalactitic. When pure, calcite is colourless and usually transparent, but when mixed with iron or other mineral colouring matter it commonly assumes yellow and brown tints.

Aragonite is also a crystalline form of carbonate of lime, but is by no means as common in Nature as calcite. It crystallises in the rhombic system, which assists the mineralogist to distinguish it from the last-mentioned mineral, from which it differs also in being harder and of higher specific gravity. Aragonite may occur as globular masses, or as incrusting other substances, or in the stalactitic form. It is sometimes white, but more often yellowish, or grey, and it is not, commonly, as transparent as calcite, whilst it often possesses one to two per cent. of carbonate of strontia, or other impurity.

It is generally stated that carbonate of lime, when deposited from cold solutions, crystallizes in hexagonal

E

(calcite), and when from warm solutions, in rhombic (aragonite) forms. No doubt, on the whole, that is the case ; but we ought not to forget that many marine organisms make their hard parts of aragonite, which, under the circumstances, is certainly not obtained from warm itions. These crystalline forms of carbonate o. re both of them found in fossil shells and the like ι. lays, and in not a few instances the calcareous constituent found in the brick-earth is present almost exclusively in the fossils, which are ground up with the rest in preparing the material for the moulding machine.

When present as hard crystalline lumps or pebbles, they have been derived from the destruction of limestones, and are then the greatest nuisance imaginable to the brickmaker and the most dangerous constituent at the same time. With proper machinery these hard lumps may be ground down to fine particles, but they are even then only to be admitted into the earth on sufferance. The best plan, without doubt, is to remove them altogether from the raw earth. They are commonly met with in what the geologist calls " boulder clay "—a deposit owing its origin to glaciers and icebergs. Very often the pebbles alluded to are not crystalline, but of an earthy character, as is the case when made of chalk. In the semi-dry process of manufacture, it is next to impossible to incorporate the ground-up particles of carbonate lime sufficiently well to result in the production of such a homogeneous earth as is desirable for making a first-class brick.

In sandy clays or loams, and in a few stiff clays used for brickmaking, certain remarkable concretions called " race " are found, the deleterious properties whereof are so well known to the average brickmaker that he carefully avoids the particular strata in which they occur. It is fortunate that these concretions have a habit of being confined to narrow limits along definite horizons in the brickyard section, so that they may be readily discarded in working. But that is not always the case, and little nodules of " race " are usually more or less frequent also in the beds above and below the horizons referred to. They are composed wholly of carbonate of lime, and their general effect in the kiln, and afterwards, will presently be explained. Other forms of concretions are known as " septaria,"— tabular or rounded masses of argillaceous limestone found in practically all stiff clays. These are often of enormous size, and are disposed in regular lines which the field geologist takes to indicate bedding planes in the clay—otherwise often very difficult to make out. In certain stiff clays little pellets of the same substance are found. The larger septaria have commonly been cracked in various directions, the fissures being subsequently filled with calcite.

Coprolites are impure varieties of phosphate of lime, and the term should, properly speaking, be restricted to a substance of organic origin,—the fossilised excrement of animals. But the name is now loosely employed to designate phosphatic concretions in general, such as are commonly found in stiff clays, in certain " greensands," and in other sedimentary deposits. The dark brown phosphate of lime has formed on and often completely envelopes many fossils ; in certain cases it has in fact been utilised as a petrifying medium, in which form it ordinarily occurs in the thick black clays of Peterborough, Cambridge, the gault of Kent, Surrey, etc.

Summing up the effects of carbonates and other

kinds of lime in the kiln, it may be at once said
that when present in any other form than as extremely
minute particles, they are distinctly to be avoided. The
small pellets and large pebbles especially are to be
avoided, for the following reasons. Carbonate of lime is
made up of lime and carbonic acid ; if a lump of this
be subjected to great heat and thus calcined, the carbonic acid is driven off, escaping by means of flues, the
open chimney, or kiln. The product is lime pure and
simple—ordinary builders' lime. Everyone knows that on
the addition of water builders' lime becomes " slacked,"
and eventually, after a fashion, "sets." Precisely the
same thing occurs in the brick-kiln. The raw brick is
often composed of pieces of chalk or other limestone,
in limestone districts and in areas where boulder clays
are largely employed for brickmaking. On being subjected to the heat of the kiln these pieces are promptly
reduced to the condition of lime. During the process
of conversion considerable expansion takes place, and
subsequently contraction, leading to the formation of
cracks radiating from the fragments of limestone, the
homogeneity of the bricks being at once destroyed.
Apart from this, when placed in the open air the lime
becomes slacked, and the quality of the brick is seriously
impaired.

Lime is a highly refractory substance, strongly basic
in character, and forms fusible compounds with silica
and other acid bodies. It is, therefore, useful as a flux
in many earths used in brickmaking, being added to
them expressly for that purpose, to the general improvement of the brick. The celebrated Dinas bricks,
for instance, are composed of a highly refractory earth
containing about 97 per cent. silica, the remainder being
lime, oxide of iron, alumina, alkali and water. To

render this material fusible and so as to make refractory bricks, from 1 to 3 per cent. of lime is added.

But what we more particularly desire to draw the reader's attention to at the present stage, is not the employment of lime in making fire-bricks so much as its mixture with ordinary brick-earth, as in the manufacture of malm bricks. Sometimes the mixture has been effected by Nature, as is the case with true marls; but the brickmaker does not care so much for these, as without considerable and expensive artificial assistance they do not often make readily saleable bricks. The common practice is, briefly, to grind chalk or similar earthy limestone in the wet state, and then to introduce it to the brick-earth with which it is thoroughly incorporated; and there are many ways of doing this, which we shall not attempt to describe now. The object of adding chalk to the brick-earth is twofold; in the first place it assists in diminishing the contraction of the brick on drying, i.e., before burning; and secondly, it acts as a flux in the kiln by combining with the free silica, or the silicates, in the earth. Undoubtedly the second is, theoretically, its chief function; but its beneficial effects in that direction are largely marred by insufficient burning, whereby a large proportion of the chalk is not actively engaged, as may be seen on examining the majority of malm bricks with the microscope. Indeed, the eagerness to save fuel, and to turn out the bricks as rapidly as possible, often leads to the chalk particles being utterly useless. And, if we may judge from conversations with several brickmakers, it would seem that the real reason why the limestone is used at all is unknown to them, except that it produces bricks of a saleable colour. This question of colour is the all-predominating one with most malm brickmakers.

We said just now that the fragments of limestone in the raw brick are reduced to lime on being burnt ; some of the latter, however, as may be anticipated from our subsequent remarks, is engaged in forming a flux wherever possible in the immediate neighbourhood of such fragments : it is the " kernel " that is left which becomes " slacked," and weakens the brick. The object of utilising the smallest particles only of the carbonate of lime is thus obvious ; and if it were possible to use ordinary builders' lime instead of carbonate of lime, the result would be better still. The difficulty in utilising builders' lime is, of course, its certainty of slacking during the preparation of the brick-earth with which it would have to be thoroughly incorporated.

SELENITE.

The " petrified water " of the brickmaker. It is a crystalline form of gypsum—a hydrous sulphate of lime, occurring in large quantities in the commonest clays used in brickmaking. Large and beautiful crystals, some of them radiating from a central point, are found in the London Clay, Kimeridge Clay, Oxford Clay, &c. By expelling the water from selenite, or gypsum, plaster of Paris may be prepared. In the kiln, therefore, it is important that this constituent be as finely ground as possible, so as to localise the effects of the anhydrous sulphate on being moistened subsequently. In hard burnt bricks, no doubt, a great deal of it is effectively used as a flux to other constituents of the clay ; but in by far the larger quantity of bricks this sulphate is re- duced to fine powdery particles easily picked out as being softer and lighter in tint than the remaining constituents

The weather-resisting qualities of the brick are naturally, not improved when much baked selenite is present ;
and the colour of the whole is apt to become variegated—
that is, in a fairly soft brick.

DOLOMITE.

Dolomite is, chemically, composed of the carbonates
of lime and magnesia in about equal proportions. It is
found as rhombohedral crystals, the faces of which are
often curved ; also in granular and massive conditions.
Its prevailing colour is light yellow both in crystals and
rock masses, but, as with most other minerals, impurities
occasionally make it assume other tints, principally red
and green. Carbonate of iron is frequently present,
sometimes to such an extent as to entirely alter the
character of the substance. As separate crystals dolomite has very little interest for us, though rarely it may
take the place of calcite or aragonite in the fossils of
brick-earths and clays. But in its massive condition, as
magnesian limestone, it is of increasing importance to
the brickmaker. For many years it has been utilised in
the manufacture of basic bricks, though at the present
moment the market in these materials is attentively
looking at the possibilities of the next mineral to be
described.

MAGNESITE.

Magnesite is pure carbonate of magnesia—that is, magnesia = 47·6, and carbonic acid 52·4 per cent. It usually
occurs massive or fibrous, but sometimes granular, and
its fine rhombohedral crystals are well known. Like
dolomite, its prevailing tint is yellow or light brown, but,
when very pure, is as white as snow. It is usually

associated with serpentine rocks. In the kiln it is highly refractory, and behaves very much in the same way as lime—forming fusible compounds with silica and silicates. For the higher grades of basic bricks it is at this moment largely exploited-in the few localities where it occurs in paying quantities. A few years since, investigation to determine the best basic refractory material was actively prosecuted in Germany, and magnesia, preheated at the highest white heat, was awarded the palm. Magnesite, when calcined, yields magnesia, which, however, still contains the impurities that might have been present in the raw material. An average percentage composition of the magnesite of commerce shows it to contain magnesia 45, carbonic acid 50, lime 1·5, protoxide of iron 1·6, the remainder being silica, alumina, and protoxide of manganese. The presence of silica in magnesite is an objection, because it is liable to have a fluxing effect at high temperatures.

Magnesite has been found in paying quantities in California, Styria, and recently in Greece. In Eubœa, in the last-mentioned country, the mineral occurs in lodes which, near Krimasi, are worked on two levels 30 to 40 feet from the top, and dipping at an angle of about 70 degrees. The general average of the lode gives 88 per cent. of carbonate of magnesia, and the substance is peculiarly suitable for the manufacture of basic bricks. A novelty with the raw material is that the proprietors sell either by guaranteed degree, or degree of analysis, the former being 95 per cent. of pure magnesia, whilst the latter often gives as much as 97·8 per cent. In inferior grades the principal increase is in the proportion of silica.

SALT.

Chloride of sodium, or common salt, is present in

many natural clays, especially (in England) in that formation known to geologists as the Trias, developed largely in Cheshire. The influence of a salt-bearing bed is, naturally, not confined to the immediate vicinity of the formation; salt being so readily soluble in water, it comes forth from the rocks in springs, which, flowing over loams and other similar absorbent earths, impart a saline character to them. In this manner otherwise useful earths for brick-making are rendered absolutely unfit for the purpose. Salt is one of the most powerful fluxes known; when mixed even in very small quantities with clay it becomes impossible to make a good brick of the substance. But we must recur to this matter at a later period in another connection. The fluxing property is sometimes taken advantage of by mixing salt with sand in moulding, or in employing a sand already saline, as when dredged from the sea, or obtained between tide-marks. A species of glaze is produced on the brick by the action of such moulding sand.

We may ignore the presence of a number of minerals such as rutile, augite, and hornblende in brick-earths, as they only exist therein in such small proportion, and have no appreciable effect in the kiln.

CHAPTER VII.

THE CHEMISTRY OF BRICK-EARTHS.

Introduction: THE BLOWPIPE.

IT is not our intention to write an elementary treatise on chemistry; but we know it is the custom for brick-makers to have chemical analyses of their raw earths made, and we are aware also that the precise meaning to be attached to these analyses is very little under-stood. Our principal aim in introducing this subject, then, is to interpret, in an elementary manner, certain typical analyses of earths and substances used in brick-making; but before doing so we shall explain some easy methods of examining earths by means of the blowpipe, which will not merely give some insight into their chemical constitution, but will afford the intelligent brickmaker a means of investigation which he can himself put into practice.

The results of a chemical analysis of a compound earth, as ordinarily used by the brickmaker, widely differ from those obtained by a mineralogical or petrological examination. The petrologist views the earth as a mineral aggregate, the constituents of which may be ascertained on appeal to a properly-constructed microscope—that is, in the majority of instances. By noting the relative proportions of the different minerals, he is enabled to state, with approximate accuracy, what is the ultimate chemical composition of the whole. From this it would appear that a rough chemical analysis could be drawn up by the petrologist without having recourse to the ordinary methods of chemical investigation. And in a

limited sense that is true. But we should not lose sight
of the fact that there is, in too many cases, an amorphous residuum in earths, the nature whereof the microscope is powerless to reveal. It is upon this remnant
that the chemist should direct his most careful attention.

The mineralogist also can give a shrewd idea of the
chemical composition of a brick-earth by using a blowpipe and accessories. This, in fact, may be regarded
as a chemical means of investigation; but it possesses
this serious drawback, viz., the blowpipe only yields a
qualitative, and not a quantitative analysis. In other
words, it can tell us something concerning chemical
compounds present in an earth, but rarely informs
us as to the relative proportions of them. Even this,
however, is of great service in many instances, though it
does not possess the value of a quantitative analysis.
For example, we have stated previously that certain ingredients are very undesirable in a brick-earth, even in
minute quantities; and that fact becomes of increased
value if we extend the field to earths used in terra-cotta,
and china and porcelain manufacture. Now, the blowpipe is a handy instrument; it may be carried about
by the prospector with its usual accessories, and occupies
but little space. Suppose he discovers a bed of white earth
which he believes to be good china-clay; he can prove
that fact, or at least obtain a great deal of information to
that end, by the mere use of that useful little instrument. Knowing, for example, that fluorine is an undesirable constituent in such a clay for many high-class purposes, he might test first of all for that; iron, perhaps, may
come next, and so in a few minutes he is enabled to
arrive at some valuable particulars that would take
much longer to obtain by chemistry in the wet way.

It will be profitable, therefore, for us to briefly describe the blowpipe and the most common of its
accessories, stating results obtained in dealing with
substances frequently met with in brick-earths. With
but little practice anyone can use the instrument, though,
as with most other methods of scientific investigation, it
requires expert knowledge to yield really excellent
results. The simple minerals and compounds to which
we shall direct attention may be detected with the
greatest ease.

The essential constituents of a blowpipe outfit are as
follow :—

1. Blowpipe.
2. Lamp.
3. Platinum-pointed forceps.
4. Platinum wire.
5. Charcoal.
6. Glass tubes.
7. Chemical reagents.
8. Miscellaneous articles.

1. The Blowpipe.—Common forms of blowpipe are shown
in fig. 5. A may be described as follows. It consists of

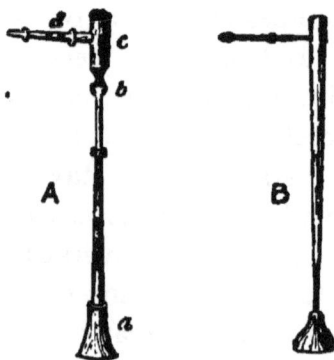

Fig. 5.—Blowpipes.

three separate parts : a tube *a b* having a mouthpiece ;
an air chamber *c* to retain moisture caused by the breath
of the person blowing ; and a side tube *d* ending in a
platinum-tipped jet. Another form of blowpipe, which,

however, does not differ essentially from that just alluded to, is shown in fig. 5, B. It is not absolutely necessary to have the jet made of or tipped with platinum, though certain examinations with the instrument are facilitated by the use of such a tip. An essential point is, that the hole in the jet should be of proper size, usually about 0·4 mm. The trumpet-shaped mouthpiece shown in the diagram may be dispensed with.

2. *The Lamp, or Candle.*—A convenient form of lamp is a Bunsen gas-burner furnished with a special jet (fig. 6, A). For certain purposes, however, this flame cannot be employed, as when testing a substance for sulphur, as coal-gas frequently contains sufficient sulphur to vitiate results. Moreover, in country districts and in the field coal-gas is not always procurable. A convenient form of lamp, though rather too large for transporting purposes, is known as Berzelius' blowpipe

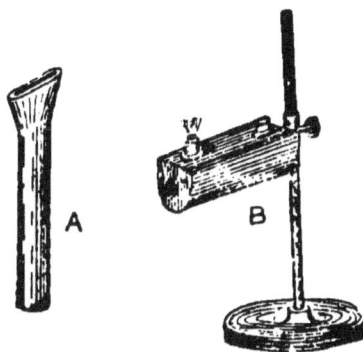

Fig. 6.—Blowpipe Lamp, &c.

lamp. This, as improved by Plattner, is shown in fig. 6 B. This consists of an oil vessel on a stand provided with two openings closed with screw-caps, the one opening being used for charging the lamp with oil, the other being fitted with a burner bearing a flat wick.

The lamp may be adjusted to any required height on the stand by means of a screw. Olive oil, or refined rape oil, is usually burnt. A spirit lamp with a flat wick is sometimes used. In countries where neither coal-gas, alcohol, nor oil are readily available, the prospector may use a small grease lamp. This consists of a cylindrical box of thin metal having a wick-holder soldered on one side, through which a flattened wick is drawn. The box may then be filled with grease, solid paraffin, old candle-ends, or fat of similar description. Professor Cole describes* it as follows :—When brought into use the wick is lighted, and the flame directed with the blowpipe upon the surface of the solid tallow or fat, until this is melted to a depth of about a quarter of an inch. The lamp will then become hot enough during use for a continuous supply to be maintained; but it is still better to hold the lamp with the pliers over a spirit lamp until all the contents become fluid. When about half or three-quarters empty, it is well to drop in extra lumps of fuel—a single candle-end or so—during use, and this additional material becomes melted up slowly with the rest. The wick must be freely supplied with fluid fuel, or it will char and waste away. If the lamp is kept sufficiently hot, the wick will not require raising during a day's work; but it can be easily thrust up with a knife point after the flame has been at work a few minutes. A cylindrical cap fits down upon the lamp when put aside. For many ordinary purposes a good carriage-candle may be employed to give a blowpipe flame, but candles have the disadvantage of not remaining at a constant level—an important point when one is comfortably at work.

---

* "Aids in Practical Geology," 1893, page 36.

3. Platinum-pointed Forceps.—At least one pair of forceps is needed, and it should preferably be made of steel, nickel-plated to prevent rusting. One end has platinum points self-closing by means of a spring, so that the piece of mineral to be heated, placed between them, may be firmly supported. At the other end are other forceps of ordinary pattern for picking up small fragments ; this end, however, should never be placed in the flame. A pair of common self-closing forceps might also be at hand for holding test-tubes, etc., in the flame.

4. Platinum Wire.—A few inches of thin platinum wire are indispensable, and lengths of an inch or so may be fixed into suitable handles. A convenient method is to have a small glass rod for a handle, and by fusing the tip of one end of the rod the glass may readily be made to hold the piece of wire. Pieces of platinum foil are useful, also, as will presently be seen.

5. Charcoal.—The outfit should comprise several pieces of charcoal, and a convenient form for each piece is a circular disc about an inch in diameter, flat at the top and convex beneath. Long prisms of the same material, square in section, are occasionally required ; these may be up to 6 inches, or so, in length.

6. Glass Tubes.—These should be of hard glass, small, of several diameters, the bore being large enough to place fragments of minerals or earthy substances within. Closed' tubes, such as test-tubes, are always requisite.

7. Chemical Reagents.—These are, for the most part, used as fluxes, and those most commonly employed are borax (sodium tetraborate), soda (sodium carbonate), and salt of phosphorus or microcosmic salt (phosphate o soda and ammonia). Small quantities of potassium bisulphate (in a glass bottle), as also small bottles of

hydrochloric, nitric, and sulphuric acids, and a solution
of cobalt nitrate, are also useful in certain cases. It is
hardly necessary to remark that the chemicals employed
must be of the highest degree of purity.

8. *Miscellaneous Articles.*—Strips of test paper, both
turmeric and blue litmus, a small hammer, a steel anvil
about an inch cube, a bar magnet, a pair of cutting
pliers, a three-cornered file, and a few small watch-
glasses are very desirable, though not absolutely
essential.

The reader, on glancing at the foregoing formidable
list of articles, may possibly imagine that some considerable outlay is requisite, and that they must occupy much
space. But that is not the case. An ordinary blowpipe, a grease lamp, a small spirit lamp, and all the
articles mentioned in paragraphs 3 to 8, both inclusive,
occupy but a small space. They may be packed in a
box specially fitted, and one in the writer's possession,
containing all of them, measures only 10 inches by 5 inches
by $3\frac{1}{4}$ inches, and is less than 3 lbs. in weight.

Now, as to the use of these various things. First of
all, let us examine the flame, as produced by a candle,
which is typical of flames obtained by other means
described, except the Bunsen lamp. A candle flame (see
fig. 7) consists of the following parts :—

1. A dark core (*a*), which contains the gaseous pro-
ducts of decomposition given off by the melted tallow
drawn up by the wick.

2. A highly luminous zone (*b*), in which only partial
burning of the combustible gases takes place. In this,
oxygen from the air combines chiefly with the combustible hydrogen, whilst the carbon is separated in a
highly heated state, which causes the luminosity.

3. An outer mantle of blue tint (*c*), where the oxygen

of the air is always present in excess, so that the separated carbon is here burnt. The highest temperature is found in this part of the flame.

Technically, the outermost zone (*c*) is known as the *oxidising* flame, and the inner luminous zone (*b*) the *reducing* flame. The two portions of the candle flame act in different manners on specific mineral substances, and the blowpipe operator may use either of them at will.

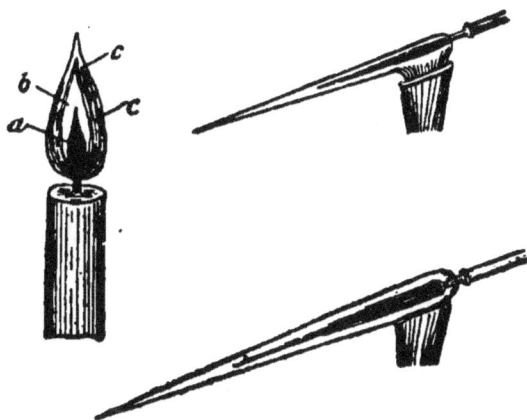

Fig. 7.—Candle and Gas Flames.

The method of doing this is illustrated in the same figure. To obtain the reducing flame, the blowpipe jet is brought to the edge of the flame a little distance above the burner, or wick. The operator then produces a gentle blast, which deflects the latter (upper figure) without altogether passing into it, so that the flame is still charged with glowing carbon. A yellowish luminous flame is the result, the most active part of which lies at a short distance from the end.

On the other hand, the oxidising flame is utilised by passing the blowpipe jet a little farther into the flame (lower figure) and blowing more strongly. A pointed non-luminous flame is the result. This will be seen to possess

F

an inner blue cone, before the point of which the hottest part is situated. Substances to be fused are placed in this part of the flame, whilst those to be oxidised are placed a little farther away, in order that they may be expcsed to the air at the time they are being highly heated.

The " platinum wire " is an absolutely indispensable adjunct to a blowpipe outfit, and is employed as follows:— A short piece of the wire, an inch or so in length, being attached to a handle, as previously described, the free end of it is bent into a loop about the size of this O. This may be heated in the flame employed, or, better still, in the flame of a spirit lamp, and, when hot enough, it may be dipped into a small quantity of the powdered borax or microcosmic salt, some of which will be found to adhere to the wire. On further heating the borax it will swell out and form a number of irregular bubbles, which (heat still being applied) will subsequently settle down into a clear, colourless bead in the loop of the platinum wire. A satisfactory bead having now been made, a portion of the mineral substance to be analysed (in the shape of small grains) is taken up by dipping the heated borax bead therein.

The actual operation of determining the nature of the substance then commences. Using the blowpipe, and directing either the *reducing* flame (R.F.), or the *oxidising* flame (O.F.), on to the substance on the borax, according to circumstances presently to be detailed, the operator notes the change in colour (if any) of the flame yielded by the process. At this point a very annoying thing sometimes happens; for, in liquefying the borax bead, it is apt to fall off the wire, and another bead has then to be made. To avoid this, great care should be taken not to blow too vigorously at first. With the microcosmic salt especial care and dexterity must be exercised in this connection. If all goes well, the pow-dered mineral substance (if fusible in the borax) readily melts down, and becomes incorporated with the borax. On permitting the latter to cool, which it very rapidly does, the bead should now be carefully examined, and any change in tint noted. Most beautiful transparent colours, pregnant with meaning, are often seen to have formed with the borax as flux.

The operator may test his skill by making the follow-ing brilliant experiments. Take up a few small frag-ments of the mineral malachite (a carbonate of copper) by means of the clear, colourless, heated borax bead, and then introduce them to the *oxidising* flame. They slowly dissolve in the borax, and, whilst doing so, the the tip of the blowpipe flame becomes emerald-green in colour. After applying this flame for a minute or two, the whole of the mineral will have become incorporated with the borax, and, when the bead is still hot, note that it is also of a rich green tint, but that, on cooling, it turns blue. If too much malachite has been taken up in the first instance, a very dark green tint is imparted, which still remains when the bead is cold, and it appears to be quite black. Its true colour, however, may be ascertained by flattening the bead out before it is quite cold. It is always well to begin by using a small quantity of the mineral substance at first, and adding to this as may be required.

Assuming that a fine rich green bead has been pro-duced, and that it contains a relatively large amount of copper, the operator may now hold it in the *reducing* flame and re-melt the bead; if the operation has been conducted carefully, the bead will then show red, and be practically opaque when cold. The red bead may now be re-heated in the *oxidising* flame, when it will be found once more to return to a green colour. The student will find this easy operation excellent practice, as proving to him, in the absence of a demonstrator, that he is really able to recognise and use the oxidising and reducing flames at will. Many mineral substances yield a distinctive colour in this way—a useful factor in a qualitative analysis.

Before using the platinum wire, be careful to ascertain that it is quite clean; a borax bead made thereon should be perfectly white and transparent.

The " platinum foil " is employed as a support during fusions ; pieces about an inch and a half long, by half an inch in width, are generally used. A small platinum spoon is sometimes adopted when fusing substances with acid, potassium, sulphate, or nitre.

Minerals may be tested to see whether, in the ordinary blowpipe flame, they are fusible or not. To do this, a fragment of the substance to be tested is held in the flame by means of the " platinum-pointed forceps." If the mineral is found to be fusible, then its " degree of fusibility " may be ascertained according to the following table. The " degrees of fusibility " are six in number :—

1. Fusible in ordinary gasflame, even in large fragments. Example: *Stibnite*, or grey antimony.

2. Fusible in fine, thin pieces, in the ordinary gasflame, and in larger fragments in the blowpipe-flame. Example: *Natrolite*, a hydrous silicate of alumina and soda.

3. If very thin splinters be used, fusible without difficulty with the blowpipe-flame. Example: *Almandite*, or iron-alumina-garnet.

4. In thin splinters fusible to a globule. Example : *Actinolite*, a non-aluminous variety of hornblende.

5. Thin edges may be fused and rounded without great difficulty. Example : *Orthoclase* felspar—already described.

6. Fusible with great difficulty on the finest edges. Example : *Bronzite*, one of the augite group of minerals.

Now, it is highly probable that many of our readers will not understand, or be able to recognise the six minerals above enumerated ; and we recommend those who may be sufficiently interested, to purchase them from a mineral dealer—such as Damon, of Weymouth, or Russell, or Gregory, or Henson, or Butler, in London. A set, comprising the six, should cost from two to three shillings. ·With these, as a standard for comparison, the operator readily grasps the method of assigning a fusible mineral to its proper degree in the scale.

Another object of examination in the forceps is to see what colour (if any) is imparted to the flame by the divers minerals experimented upon. It is a good rule not to permit the specimen, when being fused, to touch the forceps in the neighbourhood of the actual part fused. For a mineral containing antimony or arsenic would tend to form a fusible alloy with the platinum points, and so ruin the forceps.

The pieces of "charcoal" alluded to in our inventory, are used for placing the mineral substance upon in certain parts of the blowpipe operation, which may be briefly described. Essentially the charcoal forms a support to the substance during fusion ; but the glowing carbon has also a kind of reducing effect. Taking a long prism of charcoal, such as that described, page 63 *ante*, the mineral to be dealt with should be placed near one end of a flat surface and the prism so held that the flame from the blowpipe, will sweep down its full length. The object of so doing is to give a chance to any volatile substance (derived by the operation from the mineral) to deposit on the comparatively cool surface, which deposit is often indicative of the chemical nature of the mineral. To carry this point home, the following experiments may be conducted by the student. Taking a piece of *stibnite* (sulphide of antimony), which, as we have just learnt, is a most fusible mineral, we place it on the charcoal in the manner indicated. Whilst melting, and the blowpipe flame be continued to be directed upon it after it has become fused, it will be noticed that a yellowish-white deposit is taking place on the length of charcoal; this is called a *sublimate*.

Mineral substances may also be assisted in fusing on the charcoal by using the reagents described in our list of chemicals, &c., included in a blowpipe set.

In regard to the use of the "glass tubes," it may be remarked that they are used principally for the examination of minerals which yield a volatile substance on being heated therein, and to detect the presence of water and the like. It is important to make a distinction between the closed and the open tubes. When a mineral fragment is placed in a tube, closed at one end, whatever takes place will be in presence of very little air, or oxygen; on the other hand, when the tube is open at both ends, and is inclined during the experiment, a constant stream of oxygen passes through the tube, and the mineral is being dealt with in presence of that. The employment of this oxygen makes a great deal of difference in the results obtained, as a few elementary

experiments will show. If we place a piece of sulphur
in a tube, closed at one end, and heat it gently, we notice
that a yellow coating takes place inside the tube;
but if we now employ a tube open at both ends and heat
it very slowly indeed, we notice that the sulphur goes off
as an invisible gas, and if the experiment has been properly conducted, there should hardly be a trace of the
sulphur left on the glass. A number of experiments of
a similar nature might be quoted, but enough has been
said for the present to show the utility of the tubes.

The " chemical reagents " alluded to have already
been sufficiently described to render any further discus-
sion on them unnecessary for our immediate purpose.

In regard to the " miscellaneous articles " mentioned,
it may be remarked that the test papers are employed
in the detection of certain acids and bases ; whilst a
strip of brazil-wood paper is for the detection of fluorine.
The hammer and anvil are for breaking the substance to
be tested into small fragments ; the magnet for with-
drawing particles of iron from the pulverised material ;
the three-cornered file for assisting in determining the
relative hardness of minerals, &c., &c.

In examining substances before the blowpipe, it is
highly desirable that the various operations should be
carried out in some definite order. The following has
been found convenient :—

a. In a glass tube closed at one end.
b. In an open tube.
c. On charcoal.
d. With borax and microcosmic salt.
e. As to flame colouration.
f. With other reagents.

The size of the fragment to be dealt with in an examination, depends on circumstances, but for ordinary pur-

poses a piece of the size of a small rabbit-shot will be found sufficient.

It is convenient in this place to describe a few chemical reactions without the use of the blowpipe; that will render the effects on certain minerals, presently to be mentioned, clearer to the reader.

In the first place it may be ascertained whether the mineral is soluble in water, and if so, to what extent. Then as to whether it becomes soluble in certain acids, such as hydrochloric or nitric acid.   The former acid is generally used, except for metallic sulphides, and those minerals containing heavy metals, such as lead, silver, &c.; the latter is employed for the exceptions named. Several minerals, even when in a powdered state, are hardly, if at all, affected by acids.   The results to be noted during the test with acids, commonly fall into the following three groups.*

A. The mineral may dissolve quietly with or without colouring the solution; this holds good, for example, with hematite (a variety of iron), also of many of the sulphates and phosphates.

B. There may be a bubbling off or effervescence of a gas, which gas is usually carbon dioxide; but may be hydrogen sulphide.   Chlorine may be liberated, or reddish fumes of nitrogen.

C. There may be separation of some insoluble substance as sulphur, silica, &c.

We will close this chapter by stating the behaviour under blowpipe examination of various minerals, given in preceding pages, as being common in clays and earths used in brickmaking :—

Quartz.—This is infusible, and remains undissolved,

---

* See E. S. Dana, " Minerals and How to Study Them," 1895,
p. 154.

even in a miscrocosmic salt bead; but it fuses readily
with soda, on charcoal. In the flame it splinters into
fragments, which fly off with great rapidity. It is
soluble in hydrofluoric acid. *Flint*, when pure, behaves
in a similar manner.

Orthoclase Felspar.—Fusibility, 5; flame colouration
brilliant yellow, when much sodium is present; not decomposed by hydrochloric acid. It may be distinguished from other common felspars by its high degree
of fusibility.

Oligoclase Felspar.—Gives a sodium yellow flame;
fusibility, 3.5; not decomposed by hydrochloric acid.

Biotite Mica.—With fluxes gives a strong iron reaction of yellowish red colour; decomposed in concen-
trated sulphuric acid, leaving a residue of siliceous
matter.

Muscovite Mica.—When heated in a tube closed at one
end, yields water which often gives fluorine reaction with
brazil-wood test paper by colouring it straw-yellow; it
is not decomposed by acids, and whitens and fuses only
on thin edges.

Kaolin.—Is infusible; gives off water when heated in
a closed tube; and with cobalt nitrate on charcoal, a
fine alumina reaction is obtained.

Aluminium.—On charcoal, this becomes blue with
cobalt nitrate, though if the surface is fused the reaction
is not so clear. Prof. Cole advises that the soda-residue
be dissolved in dilute hydrochloric acid, then evaporated
to dryness, re-dissolved in that acid water, filter off any
silica, and neutralise with ammonia; alumina is precipi-
tated together with any iron present. The precipitate,
if white, or nearly so, may be tested with cobalt nitrate,
and the result is a fine blue colour.

Limonite Iron.—Fusibility about 5; yellow and reddish beads; water given off in closed tube; in reducing flame magnetic residue on charcoal; soluble in hydrochloric acid after a short time.

Iron Pyrites.—Fusibility about 2; yellow and red beads; in closed tube yellow precipitate due to sulphur; magnetic after reduction on charcoal; insoluble in hydrochloric acid.

Rock Salt.—Intense yellow sodium flame; fusibility about 1; microcosmic salt with copper oxide shows strong chlorine reaction—a fine blue flame surrounding the bead when re-introduced into the flame. It is soluble in water.

Selenite (Gypsum).—Fusibility about 2·5; brilliant flame; in closed tube it becomes white and opaque and much water is given off; with soda, on charcoal, sulphur reactions are obtained; soluble in hydrochloric acid.

Calcite (Carbonate of Lime).—Flame glows very strongly; infusible; effervesces freely in cold hydrochloric acid.

Dolomite.—Flame, with hydrochloric acid, like calcite; infusible; effervesces in hot hydrochloric acid.

Magnesite.—Infusible; with cobalt nitrate a fair magnesia reaction on charcoal, *i.e.*, turns into a dull pink; effervesces in hot hydrochloric acid.

Manganese.—With borax in oxidising flame a red-violet bead is obtained, but with the reducing flame it is colourless.

The above are commonly met with in brick-earths; for other minerals and substances also found, the reader may be referred to special works dealing with blowpipe analysis.

CHAPTER VIII.

THE CHEMISTRY OF BRICK-EARTHS *(Continued)*.

In this chapter we shall fulfil our promise (*ante* p. 58) to explain in an elementary manner the precise meaning. of ordinary commercial chemical analyses of some typical earths used in brickmaking, etc. We may commence by explaining a few terms used by the chemist.

An *atom* is the smallest imaginable portion of matter, and all matter is said to consist of atoms. A *molecule* is the smallest conceivable combination of atoms, and every compound substance is ultimately built up of molecules. An *element* is a substance that has hitherto defied the efforts of the chemist to subdivide or split up. Over seventy of these elementary substances are at present known, and their number is being constantly added to. Again, by improvement in analytical methods, a so-called element may be subdivided, and thus removed from the list. The elements are classified into *metals* and *non-metals ;* and it is convenient to give each of them a symbol to save trouble in writing, and to render clearer to the reader the chemical nature of a compound body. Thus, the symbol for the element aluminium is Al; for silicon Si ; for carbon C ; for calcium Ca ; for oxygen O ; for iron Fe ; for hydrogen H ; for chlorine Cl ; and so on.

We are taught by chemistry that elements are capable of combining only in definite proportions, and that each substance possesses a definite proportion peculiar to itself. That proportion is called the *atomic weight* of the element ; or, it is the relative weight of the atom of each substance compared with that of the lightest substance known, hydrogen.

Thus, the atomic weight of hydrogen being taken as 1, it is found that an atom of chlorine is 35·5 times as heavy as that, so that the atomic weight of chlorine is said to be 35·5. Now, in spite of the enormous difference between the weight of the two elements just mentioned, they combine in the same proportions by *volume ;* and the . union is known as hydrochloric acid, or HCl.

But in certain cases elements do not combine in equal proportions ; for instance, an atom of oxygen will not combine with less than two of hydrogen. Further, with this we find that the three volumes are condensed into the space of two volumes—a very common phenomenon in the chemical combination of gases. The union of hydrogen and oxygen alluded to forms water, the chemical symbol of which is, consequently, H_2O.

Chemical affinity, or *chemical attraction*, is the force which is exerted between molecules not of the same kind. Thus, in water, which, as we have seen, is composed of hydrogen and oxygen, it is affinity which unites these elements, but it is cohesion which binds together two molecules of water. In compound bodies, cohesion and affinity operate simultaneously ; whilst in simple bodies, or elements, cohesion alone has to be considered. To affinity are due all the phenomena of combustion and of chemical combination and decomposition.

Certain gases, such as chlorine and nitrogen, and such substances as sulphur, carbon, and silicon, with many others, form *acids* in conjunction with hydrogen, or hydrogen and oxygen. These combine with greater or less facility with other elements which do not form acids, and are termed *bases*. A combination of an acid and a base is known as a *salt*. Salts the names of which end in

-*ide*, such as chloride, sulphide, etc., are combinations of a metal with a non-metal. *Monoxide* means an oxide containing one atom of oxygen ; *dioxide* one containing two atoms ; *protoxide* means the first oxide, because it is the first or lowest of the oxides of the given metal in amount of oxygen present ; the highest oxide is often known as *peroxide.* The terminations -*ous* and -*ic* are frequently used for the lower and higher oxides respectively. Examples :—

FeO, iron protoxide, or ferrous oxide.
Fe_2O_3, iron sesquioxide, or ferric oxide.
FeS_2, iron disulphide.
Sb_2S_3, antimony trisulphide.

The following symbols may be indicated as referring to compounds especially met with in brick-earths :—

CaO, lime, instead of calcium oxide.
Al_2O_3, alumina, instead of aluminium trioxide.
SiO_2, silica, instead of silicon dioxide.
Na_2O, soda, instead of sodium oxide.
K_2O, potash, instead of potassium oxide.
MgO, magnesia, instead of magnesium oxide.

In *analysing* a body, the first step consists in determining the nature of the elementary substances contained therein. That may be accomplished in the *dry way* by means of the blowpipe and accessories, as explained in the last chapter. Such an examination, as previously remarked, is known as a *qualitative* analysis. Or, it may be accomplished in the *wet way* by ordinary chemical examination. The next step is to determine the amount of the constituents present, and that is known as a *quantitative* analysis. In making a qualitative analysis, the chemist is assisted by the knowledge that certain basic substances and certain acids produce peculiar phenomena in the presence of known substances or preparations termed *reagents*.

There is a great difference between a *chemical compound* and a simple *mixture* of elements ; and it is not always easy (*e.g.*, some alloys) to say whether a substance is in the one state or the other. This distinction is well exemplified by the air we breathe. The chemist finds by analysis that the air is nearly constant in composition, containing essentially in 100 parts 76·8 by weight of nitrogen (including about 1 per cent. of the recently-discovered element, argon), and 23·2 of oxygen. Small proportions of water vapour, carbon dioxide, etc., may be ignored for our present purposes. In view of this comparatively uniform composition, the question at first arises as to whether the air is, or is not, a chemical compound ? The answer is in the negative, for, amongst other things, it can be shown that the ratio of 76·8 to 23·2 is not that of the atomic weights of the two elements present, viz., 14 : 16, nor of any simple multiples of these.

We will now quote a few analyses of well-known earths, and explain each in turn :

*Chemical Composition of China-clays.**

	Kaolin.	Kaolin average.	Sandy Kaolin.
Silica	46·32	44·60	66·68
Alumina	39·74	44·30	26·08
Iron oxide	·27	·20	1·26
Lime	·36 }		·84
Magnesia	·44 }	1·60	trace
Water	12·67	8·74	5·14

* Consult " Applications of Geology," etc., by Prof. Ansted, 1865, p. 116, *et seq.*

The kaolin alluded to in the first column is a remarkably pure material, perfectly white, and contains an enormous quantity of water. It refers to one of the finest washed china-clays in the market, and is extensively used in porcelain manufacture. It is quoted here principally to give an idea of what a really pure clay is like chemically. We notice that, in spite of its relative purity, it contains ·27 per cent. of iron oxide. This could have been well done without, from the manufacturer's standpoint, but is of course a very minute proportion. Small as it is, it must exert a slight amount of colouring influence. The lime and magnesia are present in slightly larger proportions, and a little more of either would be advantageous rather than otherwise, as assisting to flux the material. This is an earth with which practically anything may be done by judicious blending and careful preparation.

With reference to the second column, the figures do not refer to any particular clay, but they have been compiled to show the average composition of kaolins as used in the market. It will be observed that the silica and alumina are present in approximately equal proportions, which is a characteristic of fairly good china-clays. The iron oxide remains as before, but there is a larger proportion of lime and magnesia —as much as can be permitted except in a second-rate clay.

The evidence of the third column shows that the sand in the china-clay is to a large extent quartzose, and this is at the expense of the alumina. Such a material would be suitable for making a species of white fire-brick, and it might do for the commoner kinds of china-ware. The earth is really of the nature of a loam— a sandy clay. There is too much iron in it for the production of perfectly white goods. The proportion of lime might be increased to advantage.

Chemical Composition of Fire-clays from Newcastle-on-Tyne.*

	1	2	3	4	5	6	7
Silica ..	51·10	47·55	48·55	51·11	71·28	83·29	69·25
Alumina ..	31·35	29·50	30·25	30·40	17·75	8·10	17·90
Iron oxide ..	4·63	9·13	4·06	4·91	} 2·43	1·88	2·97
Lime ..	1·46	1·34	1·66	1·76			
Magnesia ..	1·54	·71	1·91	trace	2·30	2·99	} 1·30
Water, etc. ..	10·47	12·01	10·67	12·29	6·94	3·64	7·58

The reader will see at a glance that the range of variation permissible in fire-clays is very wide. These earths are all found close together, and are utilised for similar purposes, though often blended to produce desired results. It will be noticed that one of them (No. 6) contains as much as 83·29 of silica, whilst another has no more than 47·55 per cent. The range with reference to alumina is very wide also, from 8·10 per cent. (No. 6) to 31·35. The refractory character of any sample of fire-clay is determined by the proportions in which the silica and alumina are contained, and by the absence of lime, iron, and other easily fluxible substances. The proportion of iron discovered in sample No. 2 is certainly much in excess of the requirements of the material, as a fire-clay, and this no doubt is tempered by admixture, unless utilised for inferior goods. The iron oxide in the other samples is about sufficient for general purposes. The amount of lime present in all the samples constitutes a good feature; much lime cannot on any account be allowed in earths for fire-clay goods. With so much iron present, and

---

* "Industrial Resources of the Tyne, Wear and Tees," 1864, p. 204.

the fair proportions of magnesia (except in sample No. 4) these clays may be regarded as typical, with the exception of No. 6. They have been utilised for many years in the manufacture of fire-bricks and the like.

Chemical Composition of Fire-clays, from Welsh localities.

	1	2	3
Silica	50·35	56·90	54·80
Alumina	23·50	24·90	27·60
Iron oxide	10·40	2·83	2·56
Soda	1·55	3·00	2·00
Magnesia	1·45	1·07	1·00
Water, etc.	11·85	11·60	11·80

The first thing that will strike the reader on looking at these results on Welsh materials, is their uniform composition as compared with the clays from Newcastle. Yet there is as much as 10·40 per cent. of iron in sample No. 1, which cannot be a first-rate clay. Its proportion of silica to alumina is, however, excellent, and, as in sample No. 3, the amount of soda and magnesia is not excessive. The soda in sample No. 2 (which acts somewhat like lime in the kiln) taken together with the magnesia and iron in the same material, is too much for a first-class clay, and would have to be suitably modified before good results could be obtained. On the whole, it is possible that sample No. 3 would yield the best results from the chemical standpoint.

We should not forget that remarkable substance of which the well-known Dinas bricks are made. The proportion of silica present ranges from about 96 to 99 per cent., the remainder consisting principally of alumina, though traces of iron, lime, and magnesia frequently occur. There is not, of course, sufficient natural flux

G

for this " clay," so a small proportion (2·5 to 3 per cent.) of lime is added, which produces the desired effect. In other words, if we can obtain a pure siliceous sand, with hardly any lime, iron or magnesia in it, we have the material of which the better kinds of fire-bricks are made. Such sandy earths are not uncommon in the South of England, but strange to relate, they are not used for the purpose indicated.

The earths from which the superior Stourbridge bricks are made, are approximately of the following chemical composition :—Silica, 64·10; alumina, 23·15; iron oxide, 1·85; magnesia, ·95; water and loss, 10·00 per cent. It will be observed that the proportion of iron and magnesia here is very small, whilst lime is altogether absent. It is a most excellent earth for the purposes for which it is used, and the chemical results may be taken as a standard for that class of material. Another Stourbridge earth yields as much as 4·14 per cent. of iron, however, whilst its proportion of silica is lower, 51·80, and alumina higher, 30·40, which serves to remind us of the variability of even good earths used in the manufacture of fire-clay goods.

Let us now turn to the consideration of pottery clays, of which the following results may be taken as typical :—

Chemical Composition of Pottery Clays.

	1	2	3
Silica	46·38	49·44	58·07
Alumina	38·04	34·26	27·38
Iron oxide	1·04	7·74	3·30
Lime	1·20	1·48	·50
Magnesia	trace	1·94	trace
Water	13·57	5·14	10·30

Some of the chief qualifications, from a chemical point of view, of earths suitable for making pottery, is the proportion and potentiality of the colouring matters present. Where the pottery is to be glazed, that is not so important ; but with ordinary unglazed ware, colour and uniformity are two highly essential desiderata. We know that the temperature employed will modify the tint, but under similar conditions the clays alluded to in the above table will give, approximately, the following results. Sample No. 1 is typical of an excellent blue pottery clay, which burns white. It contains more alumina than is commonly met with in such materials, in which respect it differs markedly also from the fire-clays just described. The proportion of oxide of iron is very small, not sufficient to perceptibly colour the finished product, though, no doubt, on careful examination it would be seen not to be perfectly white. The latitude of the term " white " is pretty considerable with clayworkers, as the reader is probably aware.

The pottery clay (also used for bricks) referred to in the middle column, is brown in colour ; it is an ordinary kind, used primarily for black and common red ware. The proportion of iron is high, and considerable quantities of both lime and magnesia exist. As might naturally be expected of such material, it will not bear exposure to great heat, though that might be regarded as a qualification in some brick and pottery yards.

The proportion of silica is high in sample No. 3, which appertains to a common yellow clay, with, possibly, some siliceous sand in it. The amount of alumina is correspondingly low, but the iron oxide is not excessive—for a common pottery clay. It is used for the manufacture of coarse ware, and burns yellow.

The chemical composition of earths used for terra-

cotta and bricks of that substance is so variable, that
without going into each case specifically it would be
impossible to convey an adequate idea. It may be
stated generally that it is not one whit less important
to consider the composition of the raw earths for
ordinary brickmaking, than in respect of that for high-
class bricks and pottery.

An excellent earth, from the neighbourhood of Ruabon,
is of the following composition :—

Chemical Composition of Ruabon Clay.

Silica	...	...	...	63·00
Alumina	...	...	...	20·10
Sesquioxide of iron	...	...	...	4·84
Protoxide of iron	...	...	...	1·51
Potash	...	...	...	2·37
Soda	...	...	...	3·10
Combined water	...	...	...	3·54
Moisture	...	...	...	1·54

The proportion of silica in this is higher than in many
clays used for brick- or terra-cotta making, but the
alkalis, potash and soda, are in strong force, so that
any refractoriness on the part of the silica is soon sub-
dued in the kiln. The iron, also, is in abundance. The
principal colouring ingredient is the sesquioxide, and
we can quite understand the manufacturer when he
informs us that, in spite of the rich tint of the goods
produced, nothing is artificially mixed with this clay to
produce such a result. We may call attention to the
method of expressing the chemical analysis in this case,
which might be copied to advantage. In the first place,
the combined and the uncombined iron are separately
shown, or rather the degree of combination is indicated ;
and secondly, the proportion of water chemically combined is differentiated from that which has simply

soaked into the clay, though expelled, following a well-known practice of chemists, prior to commencing the analysis proper. It is of very little use giving the amount of water, unless the proportions are divided in this manner. In the result given above we learn that there is very little chance of the clay shrinking, as it only contains moisture to the extent of 1·54 per cent.; but if that had been added to the water combined, we should have had a result of 5·08 per cent., which is not nearly so clear in its meaning. We may add that the Ruabon earth referred to is utilised also in the manufacture of tesselated and encaustic tiles.

In regard to the composition of earths employed in the manufacture of the commoner kinds of bricks, we may give the following examples :—

Chemical Composition of Common Brick-earths.

	Silica.	Alumina and Iron.	Lime	Magnesia.	Manganese.	Water and Loss
Reddish-brown brick clay	52·6	30·8	3·4	2·8	1·4	9·0
Red-brick clay	50·4	24·0	2·7	1·3	—	21·6
Common brick-earth ..	33·0	11·2	39·8	6·0	—	10·0
Sandy-clay (loam) ..	60·2	24·0	2·4	1·6	—	11·8

Reviewing these results, it will be noted that the brown colouring imparted to the brick in the first-mentioned example is due, to a large extent, to the presence of manganese, a rather uncommon feature in brick-earths, except where these have resulted from the denudation of iron-producing rocks rich in manganese. It will be noticed also that the proportion of water is not high for a common earth, and it must be a fairly easy material to deal with. There seem to be some possibilities in it that might, in competent hands, lead to higher things. The amount of lime and magnesia is, however, a rather serious one for a first-class clay.

In regard to the "red-brick" clay, an essential feature is the comparative absence of lime, and it would, no doubt, make "rubbers" of an ordinary kind. Unfortunately, in the results given, the iron is not separated from the alumina, but clearly the latter is very small in amount, and the results refer to a sandy material. The proportion of water is disastrous for the employment of this earth by unskilful hands. In drying, the greatest care would have to be exercised to prevent undue shrinking, and, in any case, the earth would have to be very thoroughly incorporated to make a really serviceable brick. It is with earths of this character that the majority of brickmakers *in embryo* come to grief; they know not how to handle them successfully, and twisting, warping, cracking, and "bursting" follow as a natural consequence. It is a common and treacherous material, that could only be made to succeed by perseverance and wide experience.

The "common brick-earth," as will be seen, contains an abnormal quantity of lime, and doubtless refers to a marl, though not much alumina is shown. Malm bricks could be made from it, and the product would have to be burned at a low temperature. For bricks useful to the "jerry-builder" this earth could be strongly recommended. It was, no doubt, mainly derived from limestone rocks; and, judging from the high proportion of magnesia, probably from within a watershed composed to some extent of magnesian limestone.

The "sandy-clay" or loam is of a very common type, and produces light-red bricks. There is much in common between this and the "red-brick clay" previously referred to.

The practice resorted to in various parts of the world of making bricks from slate *débris*, although not hitherto adopted to any large extent in this country, merits some description in this place. Slates may be regarded as a highly compressed clay, the original structure of which has been materially modified by the great pressure exerted during their manufacture in Nature's laboratory. To all intents and purposes they are silicates of alumina, *plus* iron, lime, magnesia, and so on, and have, practically, the same range of variation as have ordinary clays. But during their manufacture, and subsequently, certain adventitious mineral matter has been frequently introduced, as may be gathered from the following results :—

Chemical Composition of Slates.

	I	2	3	4
Silica	60·50	60·15	48·00	50·88
Alumina	19·70	24·20	26·00	14·12
Iron (protoxide	7·83	5·83	—	9·96
,, (sesquioxide)	—	1·82	—	—
,, ...	—	—	14·00	—
Lime	1·12	—	4·00	8·72
Magnesia	2·20	—	8·00	8·67
Potash	3·18	—	—	·88
Soda	2·20	—	—	—
Alkalis (not determined) ..	—	4·28	—	—
Carbon dioxide	—	—	—	6·47
Water, &c.	3.30	3·72	—	—

Analysis No. 1 refers to a blue Welsh roofing slate of Cambrian age. It is quite certain that the large proportion of alkalis present would render this material unsuitable for brickmaking, except for the commonest kinds of bricks. The iron, again, is very large in quantity, whilst the amount of alumina is low. We could not recommend this slate for good bricks under any consideration.

Analysis No. 2 is of a dark-blue slate from Llangynog, in North Wales. The amount of iron present is high, but from the low content of alkalis this material, under proper treatment, should make fairly good bricks. The ferruginous constituent is too powerful, however, for fire-bricks to be made of this slate.

Analysis No. 3, of a purple slate from Nantlle, shows a remarkable diminution in silica and a corresponding increase in iron. Lime and magnesia being present to such an enormous extent, taken in conjunction with the iron, would render this slate absolutely useless for brickmaking. There is not a redeeming feature about it.

Analysis No. 4, which refers to a green Westmorland slate, has a low percentage of alumina and very large quantities of iron, lime, and magnesia. Only bricks of an exceedingly inferior quality could result from such material.

Summing up the general characteristics of these slates from the chemical aspect, one would say that none of them are very suitable for high-class bricks. No. 2 is the best. Several minor differences will be observed between the results quoted and those referring to ordinary brick-earths — in particular, the distribution of the alkalis. A general impression is abroad that any purple slate will do for brickmaking, and manufacturers do not yet seem to have realised that the chemical nature of slates is as variable as of brick-earths. That may account for the difficulties experienced in many cases in turning out a satisfactory material. The microscope is of much use in this connexion, however, and the practical effects of chemical analyses are not always as bad as they seem at first sight.

The remainder of this chapter will be devoted to the consideration of rarer kinds of brick-earth and other raw earths used principally in the manufacture of bricks for special purposes, or as pointing to certain anomalies. As an example of what some manufacturers can do, we may quote the chemical composition of a peculiar brick-earth employed in Zurich, in Switzerland :—

Chemical Composition of Brick-earth, Zurich.

	Yellow Clay	Blue Clay.
Carbonate of Lime	23·68	27·80
,, ,, Magnesia	––	5·70
Other carbon dioxide	2·85	1·55
Silica	42·39	38·25
Alumina	18·16	12·44
Iron oxide	3·66	·73
Lime (as silicate)	—	1·85
Magnesia	—	·15
Potash	2·14	1·54
Soda	1·27	3·05
Moisture (at 100° C.)	1·27	1·37
Water, &c., chemically combined..	3·85	4·72

Here we have two clays with the carbonates of lime and magnesia present, in one case of over 35 per cent., and in the other of over 26 per cent. Professor Lunge, of Zurich, states that the bricks made from them, if burned at the ordinary heat, say a moderate red heat, are *red*, and do not keep in the air, but crumble away very soon, as the quicklime slackens on combining with the moisture. When burned at a bright red heat, about 200° C. above the former, however, they become nearly white. The lime is then present as a ferri-alumina-

calcic silicate, which causes the red colour of the iron
oxide to disappear, and, at the same time, entirely prevents any action of the moisture, quicklime being no
longer present. We have no hesitation whatever in
saying that most British makers would look down upon
raw earths such as these from Zurich, and yet many
millions of really good bricks have been made from them
during the past twenty years, and they are especially
noted for their durability. The crux of the case is the
temperature at which the earths are burned, as the
reader has perceived.

Under the heading of " magnesia," we have said a few
words regarding basic bricks. In this country they
have been made primarily from magnesian limestone,
the chemical composition of which is shown in the following results of analyses :—

Chemical Composition of Magnesian Limestones.

	1	2	3	4
Silica	3·6	2·53	·8	—
Carbonate of lime	51·1	54·19	57·5	55·7
„ „ magnesia ..	40·2	41·37	39·4	41·6
Iron, alumina	1·8	·30	·7	·4
Water, &c.	3·3	1·61	1·6	2·3

Analysis No. 1 refers to the well-known magnesian
limestone of Bolsover.

Analysis No. 2 to that from Huddlestone.

Analysis No. 3 to that from Roach Abbey.

Analysis No. 4 to that from Park Nook.

These results were obtained by Professors Daniell
and Wheatstone in connexion with an enquiry many
years ago as to the kind of stone suitable for the erection
of the Houses of Parliament.

Regarding them generally, it may be said that they are remarkable as not containing much acid, practically the whole substance of the rocks (except No. 1) being made of the carbonates of lime and magnesia. In manufacturing bricks of such materials as these, it will be seen that the ordinary methods of brick-making would not suffice. On heating magnesian lime-stone, the carbonic acid is driven off, leaving the base behind ; it is estimated that the loss of the acid, *plus* moisture dried out, leads to its reduction in weight of from 40 to 45 per cent., and the shrinkage is from 25 to 35 per cent. If water were mixed with this material, after calcination, strong chemical reactions would result, and of such a nature as to render a coherent mass of the kind required for making bricks impossible. Seeing that water cannot be employed, crude petroleum oil, coal oil, resin oil, &c., have been employed, all of them with more or less satisfactory results. The petroleum, &c., is mixed with the lime, and when the whole is burned the oil passes off, leaving bricks of solid lime. In manufacture it is highly essential to see that the lime is well burned, and it must be fresh, and not have been exposed to a damp atmosphere. An improvement has been effected by mixing from 5 to $7\frac{1}{2}$ per cent. of burned clay, which makes the lime harder after burning. An admixture of from 3 to 5 per cent. of iron oxide also consolidates the lime, though it increases shrinkage. The bricks are commonly made, in the first instance, under hydraulic pressure.

The diatomaceous earth known as Kieselguhr, which is used in the manufacture of fire-bricks for chemical works and the like, and which, for the most part, is of German origin, has the following chemical composition :—

Chemical Composition of Kieselguhr.

Silica	83·8
Lime	·8
Magnesia	·7
Alumina	1·0
Oxide of Iron	2·1
Organic matter	4·5
Water, &c.	7·1

The reader will perceive that this earth is composed very largely of silica, though there is enough iron, &c., to flux it, at any rate, without material addition. The product is extremely light, and when properly made, Kieselguhr bricks are the lightest known. They are usually of a light yellow tint, with iron spots. The silica is not in a crystalline form, the bulk of the material being composed of the hard parts of microscopic plants known as diatoms; it is more like flint.

An earth of a similar character is found in the Isle of Skye, as previously mentioned, though that burns into a redder colour.

An infusorial earth from Tuscany is composed of silica 55, magnesia 15, water 14, alumina 12, lime 3, and iron 1 per cent. That also is made into very light bricks. The general principle underlying the method of utilising those earths of organic origin is similar to that of the Dinas bricks, though they do not always require artificial fluxing.

At Saarbrücken, in the Rhenish Province of Germany, a material known as "iron brick" is manufactured. It is made by mixing equal proportions of finely-ground red clay-slate with fine clay, and adding 5 per cent. of iron ore. This mixture is then treated with a 25 per cent. solution of sulphate of iron, together with a certain quantity of finely divided iron ore. It is then moulded and baked in a special manner. We do not intend to describe the chemical composition of the various volcanic ashes, trass, and other volcanic ejectamenta used for brickmaking on the Continent in several localities. The materials of which glass-sand bricks, slag-bricks, &c., are made have no special interest in connexion with our present subject, their composition naturally varying according to the particular kinds of " refuse " employed.

CHAPTER IX.

DRYING AND BURNING.

Of the merely mechanical aspects of the operations of drying and burning bricks, we shall say little or nothing. But there are just a few points of a more or less scientific nature that offer themselves at this juncture to which we desire to allude.

The brickmaker hardly needs to be told that if he places his bricks in the sun to dry, they, or a large percentage of them, will crack, and become practically worthless from a commercial standpoint. To dry a brick properly in the open air is a lengthy operation—too lengthy for many manufacturers, who, in consequence, have had recourse to artificial drying. Many a brickyard has had to be abandoned from the inability of the worker to produce bricks that did not crack at some period of the operation, either in the drying, or burning, or both. And several manufacturers have their particular methods of "doctoring" the raw earths to prevent cracking. These are invariably "trade secrets;" though usually of a very open and transparent character, however, to the student of the subject.

It is most curious to learn the different reasons for adding this or that ingredient to the earths to prevent the brick from cracking. One who in a district has found that the addition of a little sand is beneficial, imparts that information by degrees, either personally or through his workmen, and in time it is laid down as a general axiom that "sand will prevent cracking." Another has discovered that clay should be mixed in small quantity to produce the desired result, so he and his neighbours do that, and pity the ignorance of the " sand mixers." A third feels quite certain that crushed brick, or brick dust, is a good thing ; while a fourth will add a little lime. Now, each of these ingredients is useful in its way ; everything depends upon the class of brick-earth to be dealt with. It may happen that what will, in a measure, prevent cracking, will be a bad thing in the burning, and the art of the brickmaker is to know what to do under the varied conditions.

As a general rule, where care is exercised in the dry-ing, the cracks arise from the brick-earth being too wet or plastic in the first place, and it cannot be too well understood that, *cæteris paribus*, the wetter the earth the more liable it is to crack during drying. The contrac-tion, even when the unburnt brick is shielded, and in the open air, often proves too much for the material. Then we have that class of brick-earth composed of too much clay, and that would be improved by the addition of sand—just how much depends on the particular earth ; and there is no better method of ascertaining the quantity required than by subjecting the materials to direct practical experiment in the kiln. Where no sand is available, it frequently happens that brick-dust will answer the purpose, though this may be at the expense of homogeneity in the long run. In the semi-dry pro-cess of manufacture the initial causes of cracking are not present, the block having to contract so little that it may be taken from the press and stacked in the kiln for burn-ing. Unless the brick-earth be carefully prepared, however, the surfaces of the hard blocks produced by that process are liable to develop minute cracks. And here it may be stated that unless the clay, with brick

dust or other foreign substance, be thoroughly incorporated prior to being sent under the press, and the whole
ground very fine, it is impossible to prevent cracking
during some part of the process.

Apart from the fierce and variable drying action of
the open air, we have a fruitful source of cracks in the
indentations made by stamping the makers' name or
trade-mark upon the blocks. With bricks burnt very
hard this does not so much matter, but on the commoner
kind of materials one may often perceive minute, hair-
like cracks radiating from the indentations. We pre-
sume that in this age of advertising it is impossible to
convince many makers of that fact, yet if full justice is
to be done to the material, it will be better not to make
any sharp or deep marks on the brick.

The commoner kinds of brick-earth, as we have seen,
mostly possess gross particles, grit, pebbles, &c.; these
act as so many centres from which cracks radiate either
during the drying or burning, and apart from their influence in a chemical sense, they are apt to seriously
weaken the brick.

It is truly marvellous to see how little attention many
large makers pay to the initial drying; often the long
rows of drying blocks are left unprotected except for a
rude kind of roof placed over them. The passing
shower of rain drives in underneath, and wets the
exposed surfaces, causing the clay to swell. These
surfaces, being moister than the remaining portions of
the brick, contract at a different rate, the centre occasionally being drier than the outside. The unequal
contraction produces minute cracks even in most excellent earths.

Turning to a smaller matter, the hand-barrow coming
form the drying stacks to the kiln is unprotected, which

often means that a good brick is spoilt. Of course, we are not alluding, in this connexion, to what takes place during clamp stacking; the brick produced by such a process must take its chance. The method of stacking in the kiln or clamp is very often responsible for damage to the bricks. A common method is to build them sloping outwards, and all sorts of strains and stresses are thus set up, which have their effect in producing lines of weakness, if not of actual visible cracks.

The "London stock," if not a thing of beauty, is usually strong, and that in spite of the "breeze" which forms so many points from whence cracks radiate. We must not forget, however, that a really good London stock is, above all things, thoroughly burnt, and that is a set-off against the numerous and often wide cracks.

We will assume that the brick has been either naturally or artificially dried, that no cracks have made their appearance, and that it is properly stacked in the kiln ready for burning. Now comes a most important part of the process. It is possible that any microscopic cracks will be closed by fusion or agglutination; but it more frequently happens that in unskilled hands the kiln is responsible for many cracked and "starred" bricks. To know exactly how to introduce the heat so gradually that the bricks shall not be impaired, is an art begotten only of considerable experience. Even when dealing with one particular kind of brick-earth, the maker must be careful to notice the relative moistness of his charge, and vary the mode of procedure accordingly. Suppose the brick to be as "dry as a bone" before being put in the kiln, we shall notice a considerable amount of moisture coming out of it as soon as the fires are alight; and if the heat is applied too suddenly, the bricks are not improved—they contract unevenly and

H

too quickly, and warp. When well alight, care should be taken to keep the temperature as uniform as possible, and when sufficiently burnt it must be lowered by almost imperceptible degrees. Above all things, there should not be too great a disparity between the temperature in the kiln and the outside air when unloading. Except to those who had minutely studied this matter, such a precaution might seem superfluous; it may be that no damage caused will be visible to the naked eye, but the microscope frequently shows flaws due apparently to this cause. The manufacturer may test this for himself by heating a good, sound medium burnt brick to the temperature usually found in his kiln when unloading, and suddenly plunging it in snow. It is not, perhaps, that any one of these things is especially dangerous to the brick, but it is the combined effect of all of them trending in the same direction. We desire to be clearly understood on this point. The cracks produced may not seriously impair the strength of the brick; they may be merely superficial, and they mostly are. But they materially assist the agents of denudation in " scaling " the brick, and weathering it unevenly. To this we shall return later on.

Let us now say something concerning the superficial changes produced in bricks by burning. The most important of all is the change of colour, upon which the sale of the brick depends in ninety-nine cases out of a hundred. We said a few words on this subject when dealing with the behaviour of individual minerals in the kiln. The production of an uniform tint is the main point aimed at ; and it may be at once remarked that unless the brick-earth employed is very homogeneous, or has been most carefully prepared and thoroughly incorporated, the production of an uniform colour is impossible. In regard to the tint to be produced, it should be remembered that the temperature employed in burning is a most potent factor. It is frequently laid down that such and such a temperature will form a red brick, and another and higher temperature, a blue one. That is a most absurd notion. In a general sense the principle could be correctly applied to a limited district, and with one class of brick-earth; but it cannot be made to apply all round. There is nothing like experience in regard to a point like this. In a general way, of course, a pink, red, or blue tint may be produced from one brick-earth depending upon the temperature employed; but the bulk of brick-earths would melt and the whole kiln-full be ruined in any attempt to attain such a temperature as is used in burning a sound "Staffordshire blue." Quite a large number of bricks made in the Southern half of England, may be described as having been dried in the kiln only—they cannot be said to be burnt, except that the heat employed was enough to turn them red, or to make them piebald; the particles are not agglutinated by fusion, and, indeed, there is often no trace of the constituents having been melted. On the other hand, we have red bricks in which the constituents are distinctly agglutinated by fusion, and the whole burnt thoroughly. The brick-earth of which these latter are made, would barely turn tint—would certainly not become red—at so low a temperature as that employed in producing the red in the non-agglutinated bricks alluded to.

It is not always an easy matter in burning a red brick to obtain two kilns full of the same tint, even in the same yard. When the employment of pyrometers becomes more general, that will be considerably simplified; but it is a difficult matter to get a reliable instrument, none of the forms hitherto invented being altogether suitable. That by Professor Roberts-Austen is as good as any. Many manufacturers, we are sorry to say, place colour before everything else; they even sacrifice durability to attain a certain tint. And there is much excuse for them so long as they find a ready sale for the material. When colours are made from artificially introduced mineral matter (which is not so often the case as some appear to think) the mineral introduced is, most commonly, iron; though it will be understood, from what we have previously said, that it must be used very sparingly.

The ultimate tint assumed by the brick cannot always be judged beforehand from the colour of the brick-earth. In brickmakers' language, a red clay is one that produces a red brick, a blue clay a blue brick, and so on. For the most part, colour depends on the proportion of hydrated oxide of iron in the clay; if iron is present in an earth that contains no lime, or similar mineral substance, the colour produced in the brick at a moderate red heat will be red, and at the same temperature, with the same brick-earth, the more iron present the deeper the tint. In an ordinary brick-earth, when more than 10 per cent. of iron is present, the clay is apt to burn bluish, however, and, in certain cases, almost black. With a smaller proportion of iron, and the application of intense heat, the same tint may result, and the brick become vitrified. A brown colour may frequently be obtained when the brick-earth has from 2·75 to 4 per cent. of magnesia, or a similar proportion may be artificially added to the earth.

To obtain a white brick, so that it shall also be of excellent quality, the pure white clays of Devon and Cornwall are the best, though the so-called " white " is, in the majority of cases, a light cream colour, unless, of course, the brick is glazed. In the neighbourhood of London, a whitish brick results from a mixture of chalk (carbonate of lime) with clay or loam, and is known as a "malm." In parts of Yorkshire, white pressed bricks are manufactured from common red clay mixed with magnesian lime (made from magnesian limestone) in a slacked condition. The latter ingredient, on introduction, immediately absorbs about 40 per cent. of the moisture present in the clay.

Yellow bricks can easily be manufactured from the more impure kaolins; also from certain clays in Cambridgeshire, Huntingdonshire, Kent, &c. (gault bricks); "malms" are mostly yellow, though called white.

Laboratory experiments, many years old, show that with white clay as a basis the following tints may be obtained. Phosphates of lime of various kinds = very light blue bricks. The phosphates, mixed with a quarter by weight of alum = brighter blue bricks. A mixture of white vitriol (sulphate of lime) three-quarters, with borax one-quarter = light dirty green. Sulphur and tin oxide in equal proportions = yellow. These experiments are interesting, but the ingredients would, as a rule, be too expensive for ordinary brick manufacture. They are more applicable for the production of ornamental tiles.

A time-honoured method of producing black bricks is to make any ordinary bricks red-hot and to dip them in a cauldron of boiling coal-tar for a few seconds. It is essential that the brick should be very hot, or the black staining will rub off. A good test that the operation has been successful is, that the surface shall be dull black, not shining. And there are many other ways of obtaining different tints, the description of which would be beyond the scope of the present work.

Unless a brick is extremely well burnt it is not uniform

in colour throughout. A considerable proportion of a
" draw " is often ruined in regard to tint by the adoption
of an unsuitable form of kiln. Where the brick is
actually burned (as distinguished from being baked), the
contact of the flame from the fires is almost sure to lead
to uncertainty in that respect along the flues. Impurities in the coal, such as iron pyrite, are the chief
delinquents, and there is sure to be a certain amount of
" flash." In that, as well as in the baking method,
bricks are liable to be discoloured by the bringing out of
impurities which they themselves contain.

CHAPTER X.

THE DURABILITY OF BRICKS.

THIS is one of the most important parts of our subject, and it may be approached from several points of view. When a brick decays, its structure, for the most part, is responsible therefor. A great deal depends on whether the ingredients forming the brick are merely baked in the process of manufacture, or whether they are wholly or in part agglutinated by igneous fusion. A rough and ready plan of determining this point, in the absence of experience, is by ascertaining the porosity of the brick. Other things being equal, the absorption test is undoubtedly the best all-round method of gauging the weathering qualities of a brick. But there are certain kinds of bricks which defy that method; an imperfectly burnt one with a vitreous exterior is especially treacherous in that respect, and, indeed all "vitrified" bricks are difficult to deal with by the "absorption process." Again, a brick cracked all over, not with superficial cracks only, but with those which go far into the interior, will not yield its quality by mere immersion in water. The water, it is true, finds its way right into the brick, but, as often as not, the sides of the cracks are perfectly vitrified and almost damp proof, so that on lifting the brick out of the water the latter rolls off as though it were on "a duck's back." Yet such a brick, yielding but the merest fraction as a result of the immersion, may be utterly worthless when put into a building, because it would not be strong enough.

Then we have those bricks which are seriously affected chemically, but which seem fairly good in other respects. They also, in many cases, defy the efforts of the experimenter in regard to absorption ; though they are nevertheless easily detected as being of bad quality, by other methods. Such bricks often resist great "crushing weights," and generally bear a good character, their subsequent behaviour when put in the building to the contrary notwithstanding.

In determining the weather-resisting qualities of a brick we have the following things to consider :—

1. The chemical composition of the brick.
2. Its absorptive capacity.
3. Its minute structure.
4. Its specific gravity.
5. Its strength.

The last-mentioned property can often be inferred from a knowledge of the three preceding ones, and need not, therefore, form the subject of direct experiment. In spite of that, however, we find that the "crushing strength " is much more popular than the others. The reason, so far as brick manufacturers are concerned, is not far to seek. Architects demand that especial quality. "What is the 'crushing strength' of your bricks?" enquires the architect. And if the maker does not know, he stands a good chance of losing the order. Figures are demanded, and if the maker cannot produce a higher figure than his neighbour, woe betide him. But statistics are ever deceptive, and as applied to bricks in regard to their strength especially so.

In general, we have to consider whether the brick is strong enough for the purpose to which it is to be applied ; and that depends much more on the manner in which it is built up, than on the strength of the

individual brick. For ordinary building purposes almost any kind of brick is, *per se*, strong enough, and
a mere inspection of the specimen suffices to carry conviction as to its suitability or otherwise in that respect.
For certain structures, such as buildings to carry heavy
weights—especially moving weight—for engineering
purposes, and the like, we ought, it is true, to know a
little more. Yet the engineer would be a very poor one
who could not tell at sight whether a brick submitted
to him was fit or not for the purpose he has in view,
from the point of view of its weight-carrying properties.
In any case, however, fashion demands the "crushing
weight" in figures, and although such figures are in
general of but little practical value, they must be given.

The principal difficulty the architect and engineer
have to contend with is not lack of strength, but the
setting in of decay, and that even in bricks sometimes
of the strongest description. Unless the strength is
going to be maintained, it is of no use whatever, in a
scientific sense, to give it in the first instance.

After these few preliminary observations, it will be
well to treat the subject more systematically.

THE EFFECT OF THE ATMOSPHERE ON BRICKS.

Air is a mixture of gases; dry air consists of at least four
of them, namely, nitrogen, oxygen, carbonic acid, and
argon. Of these, by far the most abundant is nitrogen,
present to the extent of about 78 per cent., then oxygen,
20·96 per cent., argon about 1 per cent., and carbonic
acid 0·04 per cent. Extremely minute quantities of
ammonia and ozone, though practically always present,
have been omitted from the preceding results of analysis
of air.

We have been speaking of pure dry air; but the atmosphere is hardly ever of precisely the same chemical composition in two different places. By the seaside it has more ozone, and chloride of sodium is found in particular abundance. In cities, especially where large factories exist, nitric acid and sulphuric acid appear most conspicuously, and the proportion of ammonia becomes larger. In the air of streets and houses, the proportion of oxygen diminishes, whilst that of carbonic acid increases. Dr. Angus Smith has shown that very pure air should contain not less than 20·99 per cent. of oxygen, with 0·030 of carbonic acid; but he found impure air in Manchester to have only 20·21 of oxygen, whilst the proportion of carbonic acid in that city during fogs was ascertained to rise sometimes to 0·0679, and in the pit of a theatre to the very large amount of 0·2734. Although these may seem to be very small percentages, yet the total amount of carbonic acid in the atmosphere is enormous, and plays a conspicuous part in the decay of certain kinds of bricks.

Sulphuric acid is found in the air of large cities principally as a product of combustion, and is, of course, a distinct impurity. A portion of this acid is free, and a larger quantity is combined. Free sulphuric acid is very destructive to clay goods in the open; and it should be remembered that the relative abundance of this impurity depends on the precise *locale* in the city. A great deal has been said and written about the decomposition of the stone of which the Houses of Parliament are built. The air in the immediate vicinity must be highly charged with both sulphuric and nitric acid from the proximity of the busy factories on the opposite banks of the Thames in Lambeth. Had the Houses of Parliament been erected, say, in Kensington, where but few factories exist, it is conceivable that the stone would have behaved much better.

Air in itself, however, has no power to destroy bricks—the various gases, acids, chlorides, salts, solid carbon, inorganic and organic dust can do nothing by themselves. But the air is always laden with vapour, the most important of which is water vapour, which condenses into rain, hail, snow, and dew. When rain is formed, the drops of water take up minute quantities of air with its proportion of carbonic acid, sulphuric acid, or what not, and it is these acids, applied to the surface of bricks through the medium of rain and moisture generally, that are liable to do the damage if the nature and composition of the brick are favourable.

Let us assume that we have a brick composed of a goodly percentage of carbonate of lime. The carbonic acid in the rain reduces this to a bi-carbonate, which is soluble in water, and hence the surface of the brick decays, the rain water washing it away. Other things being equal, it follows that the same brick will decay most rapidly in a district where the rainfall is very great and where there is the largest proportion of these deleterious acids in the air.

Whilst speaking of the various acids which attack and destroy bricks, we must not forget those formed by the decomposition of organic matter on the surface of bricks which "vegetate." The lichens, mosses, and so forth, growing from cracks in the wall, or spread over on to the brick from the mortar, yield, on decomposition, some of the most powerful acids in existence. A brick with a "crumbly" surface affords good foothold for these plants, and when they die they give rise to the so-called humus acids—crenic and apocrenic acid—which undoubtedly do an immense amount of damage. By keeping the surface of the brick moist, the plants permit the ordinary acids in rain to do more execution than they otherwise would. Taking two bricks, one which " vegetates" and one that does not, and exposing them in the same situation, it will be found that after a smart shower of rain the surface of the former has become thoroughly soaked, and the vegetation keeps it so, completely rotting it in time; whereas the surface of the latter, exposed to the same shower, may be quite dry within an hour or two after the rain has fallen.

Returning to the subject of rainfall, which exercises such material influence on the durability of bricks, we may give a few particulars concerning the distribution of rain in this country. Speaking generally, the east coast of England is the driest part of the country, the west coast having the greatest rainfall. The annual quantity at sea-level ranges from 60 to 80 inches on the west coasts of Ireland and Scotland, to about 20 inches on the east coast of England.* In some localities, however, the fall is much greater, amounting to 154 inches on the average of six years at Seathwaite, in Borrowdale, at the height of 422 feet above the sea.

The quantities which fall in particular showers are often very great, and this aspect of rainfall also has its interest for us. About London a fall exceeding an inch in 24 hours is comparatively rare, although on August 1, 1846, 3·12 inches were collected in St. Paul's Churchyard in two hours and seventeen minutes.† On our west coasts this amount is often exceeded. On October 24, 1849, 4·37 inches were collected at Wastdale Head; June 30, 1881, 4·80 inches at Seathwaite; on April 13,

---

* R. H. Scott, " Elementary Meteorology," 1883, p. 137.
† Report of British Association for 1846, Part II., p. 17.

1878, 4·6 inches fell at Haverstock Hill, London ; and a fall of 5·36 inches was recorded from Monmouthshire on the 14th July, 1875.

Taking averages of districts, we may give the following statistics, referring, of course, to annual rainfall :—

Less than 25 inches = Essex, Suffolk, Norfolk, Cambridgeshire, Huntingdonshire, Rutland, Middlesex, and parts of Surrey, Oxfordshire, Buckinghamshire, Bedfordshire, Northamptonshire, Leicestershire, Nottinghamshire, Lincolnshire, Yorkshire, and Durham. In other words, with the exception of parts of the North and East Ridings of Yorkshire and parts of Herts. and Bucks., which have a rainfall of from 25 to 30 inches, the eastern half of England, to the east of a line drawn from Sunderland to Reading, and then eastwards to the mouth of the Thames, has only a rainfall of 25 inches, or slightly less, per annum.

Between 30 and 40 inches = Practically the whole of the south coast from Kent to Devonshire, the whole of Somerset, Wilts., and the west of England generally, with the exceptions about to be noticed.

Between 40 and 50 inches = A great part of Devon and Cornwall, the western half of Wales, with the exceptions presently to be given, a great part of Lancs., and Cumberland.

Between 50 and 75 inches = A small patch in the centre of Devon, a large strip in West Wales, and an enormous tract of country in Cumberland, Westmorland, with Lancs. and north-west Yorks.

Above 75 inches = The wettest parts of the country. A small part of Dartmoor, a region in Wales in the vicinity and to the south-east of Snowdon, and the Lake District.

With reference to statistics concerning rainfall, it

should be borne in mind that those relating to special districts, especially to hilly parts of the country, are often very deceptive, and require careful local study. A slight difference in the physical features of a locality is often sufficient to lead to considerable variation—the proximity of a conical hill rising from the plain, the sudden convergence of the two sides of a valley, or, conversely, the widening of a valley into a flat stretch of land, all materially affect the local distribution of rain. A clump of trees situated in proximity to a house will frequently be the means of a downpour that would otherwise have passed over. With winding valleys great latitude must be allowed. Then, again, the geological structure of the locality is an important factor in determining the amount of moisture delivered at a given spot. Where we find a thick clay cropping out in the bottom of a valley, with more or less porous rocks rising on either side of it, we soon ascertain that the houses on the clay receive more moisture (or the latter is distributed over a longer period) than those edifices on the hill sides in the same district.

Our readers could no doubt give us plenty of instances where in a circumscribed area their bricks have behaved very erratically—the bricks of a house in one part of the district weathering well, and in another badly. That may often be due, not only to the actual distribution of the rain, but to the manner in which the rain or dew has fallen. If an inch of rain falls in the neighbourhood in one day, that would not tend to weather the bricks so vigorously as though the fall had been spread over, say, a week.

A very important aspect of the subject is that which deals with the " efflorescence " on bricks. This appears to be greatly misunderstood, being commonly assumed to be due to one set of circumstances rather than to the conspiracy of several. There are many kinds of efflorescence, and an explanation of one of them obviously will not apply to all. The " scum " that appears on the surface of bricks is, however, to some extent bound up in the composition of the rain in the particular locality where it occurs. Examined attentively, the commoner kinds of efflorescence are seen to be minute white and yellowish-white crystals. The substance of which these are formed has been drawn out of the brick, or the mortar, or both, and rain has been the principal agent in accomplishing this work, though its power in that respect must necessarily vary according to the chemical composition and structure of the brick or mortar, as compared with the nature of impurities in the rain. If some substance were present in the rain that could readily form an alliance with an ingredient of the brick, and the union was capable of crystallising out, the surface of the brick would naturally form a convenient spot for the crystallisation to take place. To prevent it, we ought to know the composition of the air at the spot where the house is to be erected, and also the chemical and physical structure of the brick to be employed. That is rather too much to expect from the manufacturer and architect ; but there is a method—we will not say an infallible one—which may be adopted to get rid of that particular kind of scum. That method could not always be adopted, as will be seen. The bricks must be burned more thoroughly, and at a high temperature; that would lead in most cases to the active employment of practically all the ingredients of which the bricks are composed, and the impurities in the rain would, in consequence, stand less chance of successfully inducing some of them to break their allegiance. In practice, however, we believe it would be found that the high temperature requisite to bring about the result just stated would either tend to spoil the colour of the brick or partially melt it. The latter could be prevented with due care, but we are afraid the former could not be so easily dealt with, with the majority of brick-earths. And if the brick is to be permanently discoloured to prevent efflorescence, it is better to permit the latter to manifest itself. The life of the "scum" is very variable; sometimes, after having once appeared and disappeared, it will never come again. The passing shower may wash it off (though it is not always so easily removed), and it may come again and again for years. It behaves very erratically. The amount of the efflorescence may be such as, in course of time, to lead to the surface of the brick "bursting" and peeling off, or, on the other hand, it may be a mere film.

There is one thing in connexion with efflorescence which cannot be overlooked in regarding its practical effects in the building. In ever so many cases we find that the scum, or the major part of it, is only to be found in the neighbourhood of the mortar joints. That is a matter of direct observation, and we have taken some considerable trouble to verify it, as it has always been regarded as a point whereon to hinge a debate. We do not say that in all cases the efflorescence appears only in the position on the brick just indicated; but it unquestionably does so in too many instances to enable us to regard its occurrence as mere accident. Taking a large surface of brickwork just commencing to show efflorescence, we find that the vicinity of the mortar joints are the first places, in very many instances, where the nuisance begins to manifest itself. From thence it spreads over the surface of the brick until the whole is more or less discoloured.

It seems impossible to deny that the mortar is guilty, to some extent, in such cases. At the same time, we must confess that we have never seen the efflorescence spreading over the mortar. It would appear that something in the mortar enters into chemical alliance with certain ingredients of the brick, and that neither without the other could produce the phenomenon alluded to. The remedy suggesting itself most readily is to chemically analyse the efflorescence, the brick, and the mortar ; supplementing the experiments with a micro-examination to see how far it is possible to locate the deleterious substances found to exist, so that they may be removed in the manufacture of the materials, if that is possible. But information on that head is of the scantiest description, and much more will have to be done before the question is definitely settled.

Another kind of " efflorescence " that often appears on bricks in damp situations is mere vegetable growth, which bears a superficial resemblance to the crystalline "scum" just described, though it can, of course, be easily differentiated on examination with a lens. The damp atmosphere is no doubt largely responsible for this, though ineffectual damp-courses are contributors. The remedy lies in having a less absorbent brick—one that will not afford ready foothold to the vegetation.

The influence of rain on the weathering of bricks may be considered from yet another standpoint. Where the brick is fairly porous, its durability is liable to be materially influenced through the agency of successive frosts. The water finds its way a short distance into the brick and saturates it. During frost the water is turned into ice at and near the surface of the brick. In forming, the ice exerts considerable expansive force, which forces asunder the particles (sand-grains and the like) of which the brick is composed—that is to say, near the surface of the brick. The accumulated effects of successive frosts in this way tends to weather the brick by breaking up its exposed surfaces. To be materially affected, however, the brick would have to be of very poor quality, and it will be seen that the presence of cracks would much facilitate the operation.

The style of a building, the manner of its construction, and especially the class of metals used for exterior decoration, all assist rain in its work. A projecting course will have its upper surface washed clean, whilst the underside remains very dirty—in cities, becoming quite black. The limit of this dark discolouration is often frayed out by the irregular action of the rain dripping from the projecting ledge, assisted by the wind. Where the projection is so designed that the rain is induced to drain to one point, and then to fall over on to the wall, an unsightly streak down the latter is the result. The free use of metal ornaments, railings, for supporting signs, for down-pipes, &c., is unfortunate in not a few instances. At the point of junction between the metallic substance and the brick into which it is inserted, or in the immediate neighbourhood above which it is fastened, the brickwork is sure to be discoloured. This may arise from the dripping of rain-water from the metal, or it may be from the decomposition of the latter, or from both. Iron rust leads to brown streaks, zinc-compo. to dirty red, and so on.

The action of the wind as affecting the durability of bricks is sufficiently important to warrant passing allusion. It drives rain and its deleterious acids farther into the brick than the moisture would soak in the ordinary way. It leads to wet walls interiorly, unless the latter are so constructed as to overcome the effects. On the other hand, a gentle breeze dries moisture on the face of the brickwork. In cities, wind indirectly assists rain and its impurities by blowing organic matter from the streets into niches and corners, where it lodges, and, decomposing, provides powerful acids capable of doing much work. Discolouration is the chief effect produced on the average brick through this medium. In certain countries, wind, by driving dust, sand, &c., acts as a species of sand blast.

Considerable diurnal variations in temperature are known to be peculiarly destructive to certain kinds of brick and terra-cotta work. Very porous bricks are not much affected, but the more compact kinds, and especially terra-cotta blocks, often suffer. These observations do not so much apply to our own country as to warmer climates ; though we are not altogether without experience here. On being heated these materials expand ; when made loosely, as in rubbers and the like, the effect of the expansion is not very manifest, because the motion is absorbed, so to speak, by the brick itself. On the other hand, increased compactness of the particles leads to a perceptible increase in the size of the bricks, and when the sun has gone down contraction takes place as the bricks are cooling. It often happens in hot climates that the brick or terra-cotta block is unable to part with its heat as rapidly as the surrounding air becomes cooler, although it tries hard to do so, and this leads to corners of the brick being broken off, the physical forces exerted during the struggle doing the damage.

A highly interesting case of the effects of temperature on terra-cotta was detailed by Mr. T. Mellard Reade, C.E., F.G.S., a few years ago.* He shews that the

* Geological Magazine, N.S., Dec. III., Vol. V, 1888, pp. 26 *et seq.*

cumulative effect of small, but repeated changes of temperature is very striking, and describes the lengthening of a terra-cotta coping in that connexion.   The coping in question, which was freely exposed to the direct rays of the sun, consisted of two courses of red Ruabon terracotta bricks set in cement upon a fence wall, built with common bricks in mortar, a brick and a half in thickness. The courses were level, but, in consequence of the inclination of the road, the coping stepped down at intervals, so that the undercourse of bricks of one length was just gripped and held in position by the top course of the next length of coping.   It will be observed that that form of construction constituted, by liability to lifting, a more delicate test than ordinarily of any increase of length, that might take place in the coping.   On subsequent examination of the coping, the end position of one length, abutting against the next length at the drop in the level, was found to be thrown up into an arch-shape bend of about 6 feet span ; the coping bricks being lifted in the highest part one inch from their bed.   There was a fracture at the crown of the arch, and another at the foot or springing, but for a distance of 30 feet the coping was practically one solid continuous bar.   A careful examination shewed that the coping had " grown " about a quarter of an inch longer than when it was first set, and that this lengthening, as shewn by movement on the corbel bricks which occur at intervals, was evenly distributed along a length of 30 feet.

Mr. Mellard Reade tells us that this is by no means an isolated case.   In the neighbourhood of Blundellsands inspection of brick copings shewed that it was quite a common feature, and he has noted several instances in which the end brickwork and piers have been badly fractured by the force of expansion.   In a case where the coping was of blue Staffordshire bricks, the top course in cement and the under course in mortar, a change in length was clearly shewn by the coping being lifted off the wall at each of the two ramps which exist in its length, and the movement was readily measured on the corbel bricks as in the case previously detailed. In this case the lengthening was also a quarter of an inch, and was evenly distributed over a considerable length of coping.

Whilst speaking of changes of temperature in their effect on bricks, we may allude to the behaviour of the material in severe conflagrations. A general rule cannot be laid down, because it is customary now-a-days to use fire-bricks for ordinary building purposes which will withstand practically any heat to which they may be subjected. Leaving them out of the question, and referring to ordinary bricks, it may be said that those of an inferior class frequently become cracked all over during a fire, or, it may be, by the sudden cooling after the fire has been put out, or by the sudden lowering of the temperature in them by the continuous action of the fireman's hose. All the same, the average brick withstands heat far better than any kind of granite, or similar igneous holocrystalline rock; loosely compacted sandstones and limestones crumble up on the surface, or flake, or may be utterly destroyed when subjected to a conflagration that would not have the slightest effect on bricks.

CHAPTER XI.

THE MICRO-STRUCTURE OF BRICKS.

THE reader may be tempted to enquire, What is the use of knowing the micro-structure of a brick? We have anticipated the question to some extent in dealing with the structure of brick-earths, but it may be well to enlarge upon it here. In the first place, the study of the minute structure enables the manufacturer to ascertain whether the brick is thoroughly and homogeneously burnt. It tells him whether the materials mixed together in the earlier stages of manufacture were thoroughly incorporated or not, whereby, if need be, he can improve that part of the process. In carefully examining what the average manufacturer would call a well-burnt brick, the microscope assists us in perceiving that it is often anything but well burnt, small local patches—"tears"—of semi-vitrified matter being observed, which should not exist, of course, in a perfectly homogeneous brick. And if the brick is not homogeneous, it suffers in respect of its strength as a whole, and in the majority of cases its colour is not uniform. To arrive at the cause of this lack of uniformity is to indicate the manner in which the manufacture of the brick may be improved, and the microscope often enables us to arrive at a satisfactory solution of the problem.

From a chemical standpoint we know that a high percentage of iron in the average brick-earth is not conducive to the production of a good brick. In the same manner by "rule of thumb" we learn that a high percentage of lime prevents the manufacture of the raw material into a fire-brick, unless, indeed, we are making basic bricks. The chemist tells us also of the respective values of potash and soda. Too much iron will cause the brick to "run"; salt has a similar effect; but beyond this the chemist cannot go, except that in the broad sense he explains what unions take place to produce such results.

The microscope, on the other hand, enables one to see exactly what has taken place; the deleterious constituents are detected at their work, and careful chemical investigation teaches us what to add to the brick-earth to neutralise the effects observed; for it is only from its effects that the artificial constitution of the brick-earth can be properly regulated.

The same instrument is extremely useful in all questions concerning the relations subsisting between a brick and the glaze upon it, the cause and prevention of the cracking of the latter, and its general quality from a physical aspect. And, speaking of cracks, we may again draw attention to the influence these have on the strength and durability of the brick: many of these minute fissures cannot be seen by the naked eye. In a similar way can the microscope be made use of in the manufacture of terra-cotta and faïence. The cracking of glazes is one of the most troublesome features the high-class brick and tile manufacturer has to deal with. If the character of the surface of the brick is not suitable for "taking" the glaze, the maker knows in a moment; the trouble is where the glaze takes readily and then, some time after the operation is finished, it becomes covered with "spider-web" cracks, unsightly and considerably detracting from the value of the brick. The

cause of the cracking is commonly attributed to the
composition of the glaze, and the manner in which the
latter is allowed to cool, and no doubt a great deal is
due on both those heads.   At the same time, we know
of many instances where the same glaze being used
under similar conditions on two different surfaces of
bricks made from one and the same brick-earth, the
glaze cracks in the one case, and hardly ever in the
other.   The direction of the cracks points to their origin,
and the character of the surface is brought in guilty.
And yet the average manufacturer would not detect any
difference in the quality of the surface—he could not,
without a good lens or low power objective, perceive the
slightest discrepancy.

The ordinary glaze behaves very much like Canada
balsam with reference to surfaces on which it is laid,
and something akin to what petrologists call " perlitic "
cracks is produced in the glaze.   We can make these
cracks, and imitate the structure artificially, by suitably
distributing the Canada balsam over the surface of a
piece of ground glass, and in other ways.   That direct
relationship exists between the cracks and the grain of
the surface on which the preparation is laid, is certain,
for we may vary the distribution of the cracks by varying the grain of the surface.   An intelligent apprecia-
tion of the disposition of cracks in glazes should be the
means of preventing them altogether, and not only with
bricks, but with faïence and vitrified work generally,
the study may be best carried on by aid of the microscope.

The microscope, also, may be made use of in identi-
fying bricks in case of dispute, though its applications
in this respect are not so important as in dealing with
building stones.

Questions of durability may frequently be decided on appeal to that instrument. Take a case in which a brick is known to contain a rather high percentage of lime: if the lime were in a combined state, the quality of the brick would not be materially affected; but assuming it were not so employed, it is possible that in a short space of time the brick would be thoroughly decomposed by atmospheric agencies. The microscope tells us at a glance the state in which that and other ingredients exist, in a well-burnt brick. We draw the line at bricks intended for the " jerry " builder ; they may well be left to take care of themselves ; we allude only to high-class productions in which science may be some aid to the manufacturer.

And now as to the microscope—for we do not use an ordinary one in such investigations. The best kinds of microscope are those used by petrologists in the study of the minute structure of rocks and minerals. The reader will find these fully described in works specially devoted to the subject,* but we may say a few words thereon.

A common form of " Student's " petrological microscope, as manufactured by Swift of London, may be described as follows :—

Eye Pieces and Objectives.—These need not be expensive, clear definition being the principal object to aim at ; the objectives should be of low power, 2-inch, 1-inch and ½-inch objectives being plenty for the purpose. Unless the reader desires to follow the subject

---

* Such as " The Study of Rocks," by F. Rutley: " Aids in Practical Geology," by Prof. Grenville Cole; " " Tables for the Determination of the Rock-forming Minerals," by Prof. Lœwinson Lessing ; " Petrology for Students," by A. Harker; and especially " Microscopic Physiography of the Rockmaking Minerals," by Rosenbusch (transl. Iddings).

from a purely petrological point of view, to study the development of trichites, globulites, skeleton crystals, etc., in vitrified bricks, in such places as these latter have cooled from igneous fusion, there is no occasion to resort to higher powers. We are far from saying that the brickmaker of the present day would not derive any advantage from studying this subject in its higher aspects, for the origin of crystallization appeals strongly to the imaginative mind, and is one of the most remarkable problems that Nature offers for our investigation. But in an elementary treatise of this kind we cannot go into the matter; and, as previously remarked, low power objectives are sufficient for our present purpose. The eye-pieces should be fitted with cross-wires, the use of which will presently be explained.

The Stage.—In the instrument we are now describing this is circular with a hole in the middle, and is so arranged as to revolve horizontally on a collar about an axis, the centre of which comes exactly underneath the centre of the objective. In other words, a straight line drawn through the eye-piece down the centre of the barrel of the microscope, and passing through the objective passes through that axis. To assist in more accurately centreing than is otherwise possible (depending on the lenses) with this cheap form of instrument, a collar with adjustable screws is ordinarily affixed to the lower part of the barrel of the microscope. The stage, with suitable clips to hold the object to be examined, is graduated so that on its being revolved it is easy to ascertain the number of degrees, at any period of the revolution, through which it has been turned. Thus, it will be observed that the object revolves with the stage. A pointer is placed in a suitable position on the frame of the microscope to facilitate the observation.

The Polariscope.—This is an indispensable adjunct, for determinative purposes it is often necessary to observe the object in polarised light. Briefly, the polariscope consists of two parts—the analyser, placed in the barrel of the microscope above the objective, and the polariser, arranged underneath the revolving stage. The analyser is so fitted that it may be shot in and out of the barrel in order that the polariser alone may be used, or the latter may be removed, leaving only the analyser in position, or both may be removed to enable the object to be examined in ordinary light, either reflected or transmitted. The lower nicol * is made to revolve, and the collar in which it is fixed is broadly graduated and furnished with a pointer.

Reflector.—An ordinary reversible and adjustable reflector is arranged beneath all.

Accessories.—For the more accurate determination of minerals, a quartz wedge, a quartz plate, etc., are used by the petrologist, but the description of these is beyond the scope of the present work. For examination in reflected light it is highly desirable to have a "bull's-eye" condenser.

An ordinary microscope with a revolving stage may be readily converted to petrological purposes, though it is better to have a special instrument.

The object to be examined may be in the form of *(a)* a fragment of the brick, or (*b*) a very thin slice of the same.

The fragment may be securely clipped and held in position on the stage, the "bull's eye" condenser being brought into use to throw a strong light on the part immediately under the objective. The polarising apparatus is no use for this, and may be thrown out of gear.

---

* Consult the works on petrology previously mentioned.

A very low power should be employed. The observation may be directed towards ascertaining how far the fragments composing the brick are agglutinated, and their size may be noted. Anything like a discolouration should be specially observed, and a minute description jotted down. In bricks that have not been burnt very hard, and in those that have merely been baked, we shall often be able to detect particles of mineral matter which further investigation, after the manner presently to be described, shows are opaque. Different forms of iron, iron pyrite, fragments of clay that have merely been dried in the process of baking, and minute pieces of chalk (now converted into lime) are amongst the most prominent opaque substances met with in common bricks. These may generally be differentiated and determined at sight, and bricks thus composed are never of good quality, though the ingredients have been ground very fine, and there may be nothing superficially to find fault with. Their bad qualities are usually brought out in the weathering. A great deal may, therefore, be learned from a careful examination of fragments in this manner.

In regard to the examination of very thin slices, that is in the majority of instances the most instructive, and, if we may say so, the most interesting method of investigation, though it must always go hand in hand with the other. The slice of the brick is so thin that the bulk of the constituents is rendered transparent, or semi-transparent. The preparation of such slices* is not diffi-

---

* The mode of preparation of thin rock sections for examination by the microscope is described in much detail in the works of Mr. Rutley and Professsor Cole previously alluded to; also in " Out-lines of Field Geology," by Sir Archibald Geikie, 1882, p. 202 *et seq*.

cult, but demands some experience; those who have neither the time nor patience to make them will find it convenient to send the fragments of brick to Damon, of Weymouth, or some other first-class dealer in geological and mineralogical specimens. The price charged, per slide, is usually 1s. 6d. At the same time, the student will find it eminently to his advantage to prepare the slices himself. In the process he will learn much that escapes attention when the work is done by another.

The thin slice mounted on a slip of glass is placed on the stage of the microscope and firmly clipped, as with the fragment. The reflector is brought into position, and a beam of light thrown through the slice— the thin section is now being examined in transmitted light. At first it will be convenient to study it with the polariser and analyser thrown out of position. A certain proportion of the constituents is found to be opaque, and should be examined in reflected light, as above described. The remainder are more or less transparent, and some of the grains will, possibly, be coloured. We notice the way in which the whole of the fragments are bound together —say, by some opaque mineral such as iron—or whether they seem to be partially or wholly fused together. In the case of a vitrified brick, the latter phenomenon is most usual, and we shall find that although crystalline fragments have been melted, or partially fused, there is commonly a centre or nucleus of each fragment in its original condition remaining, which passes through insensible gradations from the crystalline to the non-crystalline, or amorphous state. This latter circumstance may be ascertained by using the polariscope. Ignoring the opaque matter adverted to, we shall then see that what was transparent in ordinary light appears, for the most part, to be opaque in polarised light. Those portions which still let the light through are truly crystalline, and by revolving the stage we notice that they frequently change tint, becoming alternately light and dark. In that brick where the particles are agglutinated by igneous fusion, we shall observe the light decreasing in intensity from the crystalline portion (forming the nucleus, as it were, of each particle) outwards, and where the crystal fragment has been melted, so as to become fused to its neighbour, the periphery, or rather what was originally the boundary of the fragment, is quite dark. Polarised light cannot pass through non-crystalline matter, and in being melted that portion of the crystal fragment had passed from the crystalline to the non-crystalline stage. It is very easy, therefore, to determine how far the fragments composing a vitrified brick have been melted down and fused together; but to observe the phenomena under the most favourable conditions, the brick must be thoroughly well-burnt, and the section taken, by preference, from near the outside surface of the brick.

In some instances, partial fusion is so well exemplified (especially in bricks from fairly pure china clay), and the brick, after being burnt, has been permitted to cool so slowly, that devitrification has set in, when we are presented with aggregates of crystallites closely resembling the " felspathic matter " of petrologists. That is a circumstance which the maker should note well, for he has burnt the brick to the best advantage, and it is not then so brittle as it might have been had more " glass " made its appearance in the section. Prolonged heat, just above the agglutinating point, has accomplished this, and the microscope here clearly shows the advantage of allowing the kiln to cool slowly, and to permit the lapse of several days in the operation.

CHAPTER XII.

THE MICRO-STRUCTURE OF BRICKS *(Continued)*.

TURNING now to the actual appearance of minerals commonly found in bricks as they are examined under the microscope, we may remind the reader, that the physical aspect of the majority of them has already been described in those chapters dealing with the " Mineral Constitution of Brick Earths " and " Minerals: their behaviour in the Kiln," and the particulars that follow may be read in conjunction with what was there said.

It will be convenient now to describe the appearance of certain well-known minerals, as they are seen (A) in reflected light and (B) in thin sections in transmitted light, whilst the latter will, be subdivided into *1* denoting the phenomena observed in ordinary light, and *2* in polarised light. To save repetition, the letters and figures will be used to denote the methods of examination as indicated.

QUARTZ.—Present in nearly all rubber-bricks, and in the vast majority of common stocks, as well as in vitrified goods and fire-bricks. In the last mentioned, the grains are usually partially agglutinated, and are extremely minute.

A. As more or less rounded, or sub-angular fragments, white and crystalline, like clear window glass.

B. 1—Clear white, often broken up by thin hair-like lines running in various directions, and rows and patches of minute specks, which, as previously remarked, have been shown to contain fluid, &c. 2—On revolving the stage of the microscope, the crystals are usually seen to present beautiful, clear transparent colours, which in characteristic sections are very vivid—red, blue, yellow, &c.

FLINT.—Found in the same class of bricks as quartz.

A. Bluish horn colour ; irregular fragments and splinters.

B. 1—Translucent ; often melted more thoroughly than quartz in hard burnt bricks ; colourless. 2—Opaque unless in some such form as chalcedony, when an extremely minute granular aspect results, becoming slightly transparent. Melted portions always opaque.

FELSPAR.—The alteration which the different kinds of felspar have undergone in a hard burnt brick, when present, render it almost impossible to recognise them specifically.

A. Milk white, or more rarely light pink ; the mineral, even when red in the raw earths, becomes white on the application of moderate heat, as in the burning of common bricks. It is often closely fractured, and but rarely powdered.

B. The characteristic parallel lines of the triclinic varieties may often be observed, especially in rubber bricks ; but great heat, such as leads to partial peripheral fusion, frequently obliterates them to a large extent, and in a well-burnt brick it is quite impossible in the majority of cases to determine whether the felspars present are triclinic or monoclinic. More particularly is this the case when the mineral has been more or less decomposed prior to its having been burnt. The bulk of the fragments of the mineral can only be alluded to in the general term " felspars," and in ordinary light these are opaque or " fleecy," whilst in polarised light minute portions may be found to be slightly birefringent. In a

K

decomposed state it forms a prominent constituent of brick-earths in the first place, and that is precisely the material which most readily agglutinates in presence of a suitable flux. Crystallites are not uncommon in the melted peripheries, as may be seen in a hard-burnt brick in ordinary light.

MICA.—In minute flakes, shining, or glistening, and commonly black, silvery or bronze-coloured.

A. Detected at once by its thin shining scales, which frequently have not suffered much in the kiln except near the outside of the brick.

B. 1—The darker micas are usually citron coloured or light brown, and unless cut parallel to the cleavage of the mineral, exhibit a number of closely-set parallel lines, the fragments being much "frayed out" and "ragged" at the edges. 2—Using one nicol only, the mineral changes from dark to light on the revolution of the stage, and is said (in common with other minerals exhibiting a similar property) to be dichroic. With both nicols in position but little further difference is noted, except that in changing tint the whole is darker. Vivid colours are not observed except in yellows and browns. Muscovite mica is often quite white and transparent.

IRON.—Common except in white bricks made from the purest china-clays.

A. Brown or reddish-brown specks; sometimes as blue-black films in fire-bricks; dull and frequently powdery in common bricks. Surrounding, film-like, grains of mineral matter of which the brick is composed. A grain of quartz, for instance, is frequently seen enveloped by a film of red iron. Other metallic iron is more lustrous and whiter than magnetite when seen in reflected light, but such unaltered particles of the mineral could only occur in a brick that had not been subjected to great heat.

B. Opaque either in 1 or 2.

IRON PYRITE only occurs as such in bricks that have not been thoroughly burnt, or in common "baked" bricks. Higher temperatures lead to the separation of the iron from the sulphur and the general incorporation of both in the agglutination of the brick during partial fusion.

A. Brassy yellow particles.

B. Opaque both in 1 and 2.

CALCITE.—Not found in burnt bricks, nor indeed in any except those that have been sun-dried, or have been subjected to very little heat. Small pellets of lime are of common occurrence in poorly-burnt bricks. In reflected light such pellets are generally of a dirty white tint ; opaque in transmitted light.

DOLOMITE.—Practically the same observations apply as to calcite, crystals of dolomite not being found except in sun-dried bricks and the like. Under the action of much heat the mineral, like calcite, is reduced to lime

SELENITE.—This is not rare in the commoner class of bricks, though the application of much heat reduces it to the state of powder. In reflected light it is found to be present as extremely minute specks or "tears" of whitish powdery plaster. Opaque, of course, in transmitted light.

The description of the micro-appearance of many other minerals which occur but rarely in bricks does not fall within the scope of the present elementary treatise ; for practical purposes they may be ignored.

CHAPTER XIII.

ABSORPTION.

The advantage of knowing the relative absorptive capacity of bricks has been stated in these pages in divers connexions. The means of arriving at the total capacity for absorption of water, as generally practised by experimenters, are very incomplete and founded on an erroneous principle. It is admitted by all that absorption is one of the very best tests as to the quality of a brick, but such tests are meaningless unless they imitate one or other or several of the influences to which the brick would be subjected on being used in the building, or other structure.

A common method is to weigh the brick when dry and then to immerse it in water for periods varying from one to three days, subsequently re-weighing it, the difference in weight between the dry and wet states being termed the brick's " absorptive capacity."

Mr. Heinrich Ries remarks* that the absorption is determined by weighing the thoroughly dry samples, immersing in clean water from 48 to 72 hours, then wiping dry and weighing again. Vitrified bricks should not show a gain in weight of over 2 per cent. There are cases where bricks of apparently good quality shew a greater absorption than this, but they have great toughness and refractory qualities. Bricks made from fire-clays which will not vitrify so easily will, naturally, show higher absorption.

---

*16th Ann. Rep. U. S. Geol. Surv. (1894-95), pt. IV., p. 532.

Again, Mr. E. S. Fickes, of Steubenville, Ohio, has recently made* a large series of valuable tests of both paving and building bricks, in which he shews the connexion between the power of absorption and the strength of the materials experimented with. Mr. Fickes' more important conclusions are :—

1. The strength of the building brick, both transverse and crushing, varies in tolerably close inverse ratio with the quantity of water absorbed in twenty-four hours. The strongest bricks absorb the least water.

2. Good building bricks absorb from 6 to 12 per cent. in 24 hours, and with no greater absorption than 12 per cent. will ordinarily show from 7,000 to 10,000 or more pounds per square inch of ultimate crushing strength.

3. Poor building bricks will absorb one-seventh to one-fourth of their weight of water in 24 hours, and average a little more than one-half the transverse and crushing strength of good bricks.

4. An immersed brick is nearly saturated in the first hour of immersion, and in the remaining 23 hours the absorption is only five-tenths to eight-tenths of 1 per cent. of its weight, as a rule.

These experiments are of much interest and are probably approximately correct ; but we venture to think that if the absorption experiments had been carried out in a different manner, the results would have been still more valuable.

Long before the publication of the results of the last mentioned series of experiments, the present writer had discovered the close connexion which subsists between the relative absorptive capacity of bricks and their strength ; a slight correction must be applied for

---

* 16th Ann. Rep. U. S. Geol. Surv. (1894-95), pt. IV., p. 539.

specific gravity.   We are not prepared to enter into this subject at any length, but it may be observed that we should not have arrived at such close results had we experimented in the same way as the American authors just quoted (or others, for the matter of that).

When you completely immerse a brick in water you prevent the escape of air to a very large extent from the pores in the interior of the brick.   An old-fashioned way of overcoming this difficulty, was to place the brick in the receiver of an air-pump and exhaust the air, subsequently immersing the brick.   This latter method certainly possessed the merit of enabling the experimenter to arrive at total absorption very rapidly, but it did not imitate natural processes any more than does the thorough immersion of the brick in water.

A writer in the *Builder* of May 25th, 1895, p. 397, experimented as follows :—The bricks were placed in water in a large vessel, on edge, supported where necessary by flat blocks, to bring the uppermost face of each brick about ¼-inch above the surface of the water. Experience had shewn that by completely immersing a brick, the air did not get an opportunity of escaping from its pores with the same facility as when one surface was left out of water.   This disability, it was found, materially impaired the results of the rate of absorption (rate, as well as total tests, being carried out).   By arranging the experiments in the manner described, there can be no doubt that each brick absorbed the maximum quantity of water possible ; at any rate, there was no water-pressure from above to retard the expulsion of the air.

The tests in the last-mentioned case extended over one week, the relative absorption being taken at intervals of 1 second, 1 minute, 30 minutes, 1 day, and at the end of the week.   It was found that English vitrified bricks absorbed from 1.16 to about 1.85 per cent. in one week; white glazed and good red and blue facing bricks from 5.31 to 10.34 per cent. in one week; wire cut facers and rubbers, with white gaults, imbibed as much as from 12.93 to 20.50 per cent. of their dry weight in one week. The rate of percolation suggested many interesting problems, not the least important being the effect of chemical decomposition in prolonged immersions, whereby after being quiescent for a few days (after taking in the water for a few hours), absorption "burst out" again and continued to the end of the week. One thing is very apparent from this, namely, that for the lower grade brick even an immersion for one week is not sufficient for practical purposes. The writer remarks, "some of the red bricks from Bracknell, being placed in the vicinity of the white gault bricks (in the water), discoloured the latter to such an extent as to disfigure them. It was not merely a surface colouration; it extended to at least $\frac{1}{4}$in. into the interior. The red colouring matter was iron, but there was not enough of it by weight dissolved to materially interfere with the experiments. This very clearly shews, however, the folly of erecting a building coursed with white and red bricks, when both are very absorbent and the red has so little hold of the iron of which it is partly composed— unsightly stains are bound to appear."

This question of the solubility of certain ingredients of bricks, has not received the attention it deserves; and closely connected with that is gradual decomposition, whereby the brick becomes more and more porous—a potent factor in its ultimate destruction.

CHAPTER XIV.

STRENGTH OF BRICKS.

A VERY great deal is known concerning the strength of bricks. In addition to the innumerable experiments carried out by public bodies, we have the results of painstaking investigation by professors in universities and colleges, and the results carried out for and published by brickmakers themselves. Yet another large series of results have been published from time to time by professional journals, and it is, indeed, to these that we must look (at any rate in Britain) for anything like detailed work. The "Minutes of Proceedings of the Institution of Civil Engineers," the "Transactions of the Royal Institute of British Architects," the "Proceedings" of several allied provincial architectural societies, the "Builder," the "British Clayworker," builders' "Price Books," and several engineering "Handbooks," have all contributed to our knowledge in regard to the strength of bricks. Of works consecrated entirely to the subject there are none—applied to British materials; but we have that excellent text-book by Professor Unwin, F.R.S., "The Testing of Materials of Construction," and the important work by Mr. David Kirkaldy, both of the greatest possible value as being the results, largely, of original work. The experiments of recent years have been made almost exclusively by Mr. David Kirkaldy at his works in Southwark; by Professor W. C. Unwin at the Central Institution of the City and Guilds of London Institute; and by the Yorkshire College, Leeds.

With such a wealth of information a whole treatise might profitably be written, but it will be understood that in a small work like the present we can only give a comparatively few results, prefaced by observations to impart a general idea.

With the strength of brickwork, it is different, and it would seem rather remarkable, at first sight, that architects and engineers, who are every day using thousands of bricks, should have been at little pains to ascertain the " safe load " which this or that brick pier or wall would carry. Experience is, of course, of great value in all work of that description ; but there is always the lurking suspicion that the engineer is making his piers too big, and that the architect is by no means running the thing close. The real reason why so little has been done to test the strength of brickwork is the difficulty in getting machines of such capacity as would crush sufficiently large masses. Small piers have been built from time to time, and bricks embedded in putty for mortar have served their purpose, but practically nothing of a really serious nature was carried out in Britain until a few months ago. The science committee of the Institute of Architects, well knowing the advantage of information as to the strength of brickwork, have partially carried out a most elaborate series of experiments, the first fruits of which have already been published, but it would be out of place to allude to them here. When the remaining brickwork shall have been built long enough at the experimental station, the final experiments will be made, and the results will, we have no doubt, be the most important contribution to our knowledge concerning the strength of brickwork that has ever been published in the kingdom.

But we must give our attention solely to the strength

of bricks. To begin with, we must deprecate the idea that experiments as at present carried out give anything like the actual strength of bricks—the results are generally either too high or too low. Neither are the results comparative, except to a limited extent. One kind of brick has a " frog " on one side, another is recessed on both sides; a third is stamped with the maker's name, or some device by way of trade mark, a fourth is as flat on all sides as may be, a fifth is pressed, a sixth is hand made, and a seventh wire-cut, and there are many other varieties of make. With such different kinds it is next to impossible to arrive at comparative data that shall be of much use for working purposes. Again, the whole brick may be subject to the experiment, or only the half-brick. The faces placed between the dies of the crushing machine may not be flat, and they are most frequently irregular. If the dies are applied to such bricks it is evident that corners will be broken off before the brick has really suffered much, and that to get the best result the faces must either be made perfectly true and parallel to each other, or some other method adopted to put matters right. That commonly employed is to place some yielding substance between the faces and the surface of the dies. Sometimes thin sheets of lead or pine wood are inserted. Professor Unwin has the faces of the brick made smooth and parallel by means of plaster of Paris, and the brick is then crushed between two pieces of millboard or between the iron pressure-plates, one plate having an arrangement to allow for any slight want of parallelism between the two surfaces of the brick applied to the plates.

Now it will be obvious, what with the difference in the shape and the various modes of experimenting, that the results are by no means comparative unless the precise facts are given ; and when they are, it is but rarely that you can find more than half-a-dozen or so kinds of bricks of each category that offer all the elements necessary for comparison. So that, with all the wealth of information, we are by no means laden with much that is of actual comparative value, and if the experiments and their results are not comparative, of what use are they ? So long as experimenters are each allowed a different method of research, and so long as makers will have partial or whole " frogs," will stamp their names or initials, or will produce plain bricks only, so long will it be impossible to arrive at the best results that are really attainable. What we want is a government testing station as they have in Germany ; or, at least, the mode of experimenting should be under some central control. The experimenter, further, should select the samples to be crushed, and should be at liberty to publish all results obtained. At present, if the brickmaker does not like the results arrived at, he, of course, does not publish them. And, if he has had a number of experiments carried out from time to time, he will, usually, quote only the highest results on his bricks. That is perfectly natural, and would be understood as " business." All brickmakers may not do that, and a few may publish every or average results (we do not mean of one set of experiments, on say six bricks) of different experiments, but we fancy they are very rare. Therefore, in a matter so important to the architect and the engineer, and indeed to the general public, from the point of view of safety, we maintain that the whole thing should be carried out under some central control, as on the continent.

And now to proceed with the description of results on

a few typical bricks. Glancing at table 1, we may say that the strength of bricks as a whole is often quoted as here given, and has done duty for many years as the average strength of bricks. These bricks were crushed

STRENGTH OF BRICKS.—I.

Description.	Pressure in tons to	
	Crack.	Crush.
Four white bricks, each	16.25	41.00
Three ,, ,, ,, ...	17.05	41.05
Red bricks, ordinary	13.00	26.25
Red bricks, not well burned ..	13.75	25.05
Best Paviours	14.00	23.00
Grey Stocks, London	12.00	14.00

in a Clayton machine, and all were bedded upon a thickness of felt and laid upon an iron faced plate, and the experiments were conducted by the Metropolitan Board of Works.

Turning to the second table, compiled for the most part from brickmakers' circulars, and from the original results obtained for the late Building Exhibition, at the Agricultural Hall, all the experiments, we believe, having been carried out by Mr. David Kirkaldy, it will be noted that great variation in strength is apparent, following the different kinds of bricks. The highest result, 1064.2 tons per square foot, was obtained on a blue Staffordshire brick, though that is very closely run by bricks made from slate débris (1056.2 tons) from South Wales. The lowest result, 139.5 tons per square foot, was from a Worcester brick.

Table III. is by Professor Unwin,* and records the strength of several well-known bricks. Professor Unwin's mode of experimenting we have already alluded to.

* "Testing of Materials of Construction," 1888, p. 438.

STRENGTH OF BRICKS.—II.

Locality.	Description.	Dimensions, Inches.	Mean stress of six samples in tons per squ're ft	
			Crack'd	Crush'd
West Bromwich	Blue	2.74, 9.03 x 4.36	548.6	1064.2
,, ,,	Blue (another make)	2.80, 8.75 x 4.12	260.7	651.0
,, ,,	White glazed, "Terra Metallic," recessed both sides	3.10, 8.80 x 4.22 3.16, 8.70 x 4.34	225.0	273.7
,, ,,	Blue vitrified	2.55, 9.03 x 4.30	245.1	654.9
Worcester	"Pressed," recessed top and bottom	3.20, 9.14 x 4.50	65.0	139.5
,,	"Builders," recessed top and bottom	3.20, 9.30 x 4.50	56.1	155.5
Saltley, Birmingham	Red, recessed one side	3.20, 8.90 x 4.35 3.25, 8.95 x 4.40	138.7	180.5
Rowley Regis, Staffs.	Blue vitrified no recess	2.85, 8.75 x 4.20	385.6	722.7
Leicester	Red, recessed both sides	2.65, 8.90 x 4.25 2.75, 9.10 x 4.36	105.9	150.6
Napton-on-the-Hill, Rugby ..	Light brown, wire cut	2.85, 8.92 x 4.20 2.90, 9.10 x 4.25	131.6	303.9
Ruabon	Red, no recess	3.10, 8.75 x 4.28 3.15, 8.73 x 4.29	439.2	676.8
,, ...	Blue, no recess	3.02, 8.99 x 4.37 3.01, 8.95 x 4.36	358.9	561.2
Glogue, Whitland, S. Wales	Slate débris	2.33, 8.70 x 4.25	556.4	1056.2
Ravenhead, St. Helens, Lancs. ..	Red, brown wire cut	2.90, 9.00 x 4.20 2.90, 8.90 x 4.27	215.8	354.7
Earith, St. Ives, Hunts.	Yellow, wire cut	2.50, 8.70 x 4.10 2.50, 8.80 x 4.20	135.9	178.8
Gillingham, Dorset	Red, wire cut	2.60, 8.90 x 4.30 2.60, 8.90 x 4.25	159.5	261.7
Newton Abbot, Devon	Vitrified "granite"	2.80, 8.90 x 4.35 2.80, 9.10 x 4.55	..	445.2

Table No. III. is specially instructive as indicating the relative strength of several well-known bricks, the experiments being carried out solely for scientific pur: poses. Yet the figures must not be taken too seriously. Glancing at those relating to " London Stocks," we find

STRENGTH OF BRICKS.—III.

Description.	Dimensions. Inches.	Cracked, at tons per sq. ft.	Crushed at tons per sq. ft.	Colour.	Remarks.
London stock ..	4.6 x 4.1 x 2.4	128	177	Yellow	Half brick
" " ..	4.6x4.0x2.45	133	181	"	"
" " ..	9.2 x 4.1 x 2.8	—	129	"	
" " ..	8.9 x 4.2 x 2.3	—.	113	"	
" " ..	8.9x4.25x2.5	—	103	"	
Aylesford, common	8.9 x 4.4 x 2.7	48	183	Pink	
" "	8.9 x 4.4 x 2.7	111	228	"	
" pressed	9.1 x 4.3 x 2.7	71	141	Red	Deep frog
Rugby, common ..	9.5 x 4.2 x 2.9	158	190	"	{Between pine bds.}
" " ..	9.0 x 4.2 x 3.0	—	120	"	
Lodge Colliery, Notts	9.0 x 4.2 x 3.4	127	159	"	
" " ..	9.0x4.2x3.25	55	122	"	
Digby Colliery, Notts	9.3x4.1x3.25	248	[353]	"	Not crushed
" " ..	4.6 x 4.2 x 3.2	414	414	"	Half brick
Ruabon, pressed ..	8.8 x 4.3 x 2.7	361	[361]	"	Not crushed
Grantham, wire cut	9.2 x 4.4 x 3.2	—	83	"	
Leicester, " "	4.4 x 4.1 x 2.6	251	337	Pale red	Half brick
" " "	4.3 x 4.1 x 2.6	109	308	"	"
" " "	9.06x4.2x2.8	115	229	"	
Cranleigh, pressed	4.7 x 4.6 x 2.5	149	181	"	Half brick frog.
" "	4.6 x 4.6 x 2.5	165	237	"	" " "
Candy, pressed ..	8.8 x 4.3 x 2.8	80	381	—	
Gault, wire cut ..	8.7 x 4.1 x 3.0	111	173	White	
" " ..	4.4 x 4.2 x 2.5	119	145	"	Half brick
" " ..	8.7 x 4.1 x 2.9	—	169	"	
Staffordshire blue, common ..	4.5 x 4.3 x 3.0	216	464	Blue	"
" " "	4.3 x 4.2 x 3.0	152	386	"	"
" " "	8.9 x 4.3 x 3.1	240	[353]	"	Not crushed
Staffordshire blue, pressed ..	9.0 x 4.3 x 3.1	—	275	"	
Glazed brick ..	8.8 x 4.4 x 3.3	69	166	—	Frog.
" " ..	8.9 x 4.4 x 2.9	166	174	—	

the strength varied from 103 tons per square foot to 181 tons. But more recent experiments made by Professor Unwin* on some London Stocks from Sittingbourne, in Kent, shewed that with four samples one crushed at

* *British Clayworker*, April, 1896, Supplement, p. iv.

60·76 tons per square foot and another gave out 94·6 tons, the mean strength of the four yielding 84·27 tons per square foot. With such heterogeneous materials as London Stocks, we ought not to be surprised at these results, but they form a striking commentary on the value of general statements concerning the strength of bricks of varied character going by the same name in the market.

When we consider the strength of homogeneous bricks, and especially where these latter are made of thick marine clays, or where the relative proportions of earths employed are carefully attended to in the raw material, the results appear to be more generally applicable—as far as they go.

With ordinary Gault bricks we find a range in strength from 145 tons to 173 tons per square foot ; but Professor Unwin,* in his more recent experiments, finds that of four Gault bricks, one reached as high as 197·6 tons per square foot, and he gives 182·2 tons as the average strength.

To shew the absurdity of alluding to the strength of " blue Staffordshire " bricks, without also giving the precise locale of the samples dealt with, the reader is requested to refer to Table III., where the figures indicate a range from 275 tons to 464 tons per square foot, and to compare them with the results on Staffordshire bricks as stated in Table II., where we find a range from 651 tons to 1,064·2 tons per square foot. Of what value can a single formula be which gives the strength of Staffordshire bricks as a whole as based on such widely divergent figures as these ? Professor Unwin, in his recent series of experiments alluded to, finds that with four Staffordshire blue bricks, the

---

* *Op. cit.* p. iv.

weakest gave a result of 564·8 tons per square foot, and the strongest 788 tons ; the mean of the four being 701·1 tons per square foot.

The results on the Leicester " reds " are no more encouraging; the figures in the foregoing tables are 150·6 tons, 229 tons, 308 tons, and 337 tons per square foot. Similarly, Professor Unwin has more recently found that the Leicester " reds " from Elliston, near Leicester, bear a crushing strain varying from 311·4 tons to 591·4 tons per square foot in four samples.

From the foregoing it will appear to the reader that average results are of very little value to the architect or engineer, unless—(1) the brickyard is mentioned from which the bricks experimented with came ; (2) the particular class of brick from that yard ; (3) the method of experimenting, as to whether any substance was placed between the dies of the press and the brick to be crushed, and if so, what ; (4) if recessed or initialled ; (5) whether machine or hand made, and (6) as to whether the surfaces of the bricks were concave, convex, or flat.

Results on bricks not localised are not of much value, and it is absolutely useless for working purposes to give in one figure the strength of " London Stocks," " Staffordshire blues," " Leicestershire reds," and the like. In a general way, of course, it will be admitted that the " Staffordshire blue " is a stronger brick than the " London Stock," and so forth ; but that is as much as can be permitted—it is of no practical use to give relative figures in general terms.

It frequently happens that the capacity of the machine used for testing the strength of bricks is not enough for those bricks having a very high resistance to crushing. In the recent experiments by Professor Unwin, more than once alluded to in this article, it was found necessary to experiment with half-bricks only, and he ascertained that bricks tested as half-bricks shew about 25 per cent. less resistance per square foot than when tested as whole bricks.

Further observations on strength are made under the next heading in connexion with other forms of testing the value and physical properties of bricks.

CHAPTER XV.

ABRASION, SPECIFIC GRAVITY.

Abrasion.—In this country it is not customary to test bricks and stone by means of the abrasion process, though many English materials have been dealt with in this manner on the continent.

Abrasion tests are of special value in regard to paving bricks, and this mode of experiment is largely carried out in the United States. As Mr. H. Ries remarks,* the abrasion test approximates closely the conditions under which the paving brick is used, and is, therefore, an important one. The usual method of conducting this test is to put the bricks in an ordinary "foundry rattler," filling it about one-third full. It is then rotated at the rate of about 30 revolutions per minute, and about 1,000 turns are sufficient. The bricks are weighed before and after to determine loss by abrasion.

A more recent modification is to line the "rattler" with the bricks to be tested and then put in loose scrap iron. This is claimed to give more accurate results, and avoids loss by chipping due to the bricks knocking against each other, as in the previous method, although that has been somewhat obviated by Professor Orton, jun., by the introduction of a few billets of wood into the rattler.

The abrasion test may also be made by putting the weighed bricks on a grinding table covered with sand

---

* 15th Ann. Rep. U.S. Geol. Surv. Pt. IV., 1895, p. 532.

and water, and noting the weight before and after grinding. This last method seems to us to be decidedly the best, provided the bricks be weighted, that the weight is constant, that the feed of sand and water is uniform, and that the bricks to be tested are placed equidistant from the centre of the turning table. If this last point be not attended to, it will be obvious that in course of the revolutions the sand will tend to accumulate towards the centre of the table, and the bricks placed in that vicinity would receive more than their fair share of abrasion, as compared with those bricks situated near the edge of the table. Conversely, those bricks near the periphery would be subjected to greater grinding action, from the circumstance that the table would move faster underneath them than under those bricks nearer the centre of the table.

The bricks should certainly be weighted in such abrasion tests, and it seems desirable that the weights should be so adjusted that the weight of the brick is also taken into account. It is obvious that the abrading action of, say, street traffic, will be the same on a brick, no matter what the latter weighs, depending on the area of surface exposed to traffic. And if we experiment with one brick, weighing say 7 lbs., and another weighing 14 lbs., the greater weight of the latter, will (*cæteris paribus*), by the abrasion tests as usually adopted, give a much higher result than would the lighter brick. On the other hand, if the 7 lbs. brick be weighted another 7 lbs., then the results would be strictly comparable, provided always that the area exposed to abrasion in each case be the same, and that the other conditions we have laid down are strictly observed.

Knowing as we do that the rough and ready method of " rattling " cannot possibly give truly comparative

results, we do not intend to enlarge much on the results
of the American tests ; but the following are suggestive
as shewing the connexion between the tests for absorption, rattling, and strength combined.

Some valuable and interesting tests were recently made
by the Ohio Geological Survey, to determine the relative
merits of fire-clays and shales for the manufacture of
paving bricks, as well as the influence, if any, of the
method of manufacture adopted. Twenty-two varieties
of shale bricks, or bricks the largest constituent of
which is shale, were grouped together : fifteen varieties
of fire-clay brick ; four varieties composed of shale and
fire-clay mixed in equal proportion ; and three varieties
made from Ohio River sedimentary clays. The averages
of these four classes of results were as follow :—

TESTS OF FIRE-CLAY AND SHALES.—PAVING BRICKS.

			Absorp-tion.	Rattling.	Crushing.	
					Square Inches.	Cubic Inches.
Shales ..	..	..	1.17	17.61	7,307	1,764
Fire-clay	..	..	1.62	17.32	6,876	1,673
Mixture..	..	..	1.44	18.72	5,788	1,400
River Clay	..	..	1.36	19.02	4,605	1,176

From a series of tests recently made by Mr. Fickes,[*]
the following factors were educed :—

1. A brick which stands the " rattling " test well, has
ample crushing strength and rarely chips under less than
5,000 lbs. per square inch, or crushes under less than
10,000 lbs. The crushing strength tends to vary with
the resistance to abrasion, however, but more slowly
and irregularly.

2. The transverse strength also tends to vary with the resistance to abrasion, but more slowly and irregularly.

3. The toughest bricks usually absorb the least water.

Specific Gravity.—The practical value of knowing the specific gravity of a brick has, perhaps, been a little over-rated by writers on the subject. At the same time we do not deny that there is some use in ascertaining this property. Foremost, we have to mention its value in conjunction with absorption in arriving at a rough and ready means of gauging the strength of a brick, without having actual recourse to the crushing machine. It appears to us, however, that the specific gravity of bricks is rarely quoted in a proper manner, and until there is one uniform method, the results will always be at a discount. We allude to the fact that some experimenters take the specific gravity of a porous brick, without stating whether the amount of water absorbed, during the process, was taken into account in arriving at the specific gravity or not. Theoretically, of course, the substance to be dealt with is non-porous, and experimenters, worthy the name, either render the brick waterproof, or, ascertaining the amount of water the brick has absorbed, take that into consideration in calculating results.

The writer is in the habit of quoting the specific gravity in two ways, viz. : (*a*) the true specific gravity, and (*b*) the specific gravity of the particles. In an elementary treatise like the present, however, it is not desirable to enlarge on this subject.

THE END.

INDEX, &c.